SALVADOR LURIA

SALVADOR LURIA

An Immigrant Biologist in Cold War America

Rena Selya

The MIT Press
Cambridge, Massachusetts
London, England

The MIT Press would like to thank the anonymous peer reviewers who provided comments on drafts of this book. The generous work of academic experts is essential for establishing the authority and quality of our publications. We acknowledge with gratitude the contributions of these otherwise uncredited readers.

This book was set in Stone Serif by Westchester Publishing Services. Printed and bound in the United States of America.

Library of Congress Cataloging-in-Publication Data

Selya, Rena, author.
Title: Salvador Luria : an immigrant biologist in Cold War America / Rena Selya.
Description: Cambridge, Massachusetts : The MIT Press, [2022] |
    Includes bibliographical references and index.
Identifiers: LCCN 2022000763 (print) | LCCN 2022000764 (ebook) |
    ISBN 9780262046466 (hardcover) | ISBN 9780262368346 (epub) |
    ISBN 9780262368353 (pdf)
Subjects: LCSH: Luria, S. E. (Salvador Edward), 1912–1991. | Microbiologists—
    United States—Biography. | Microbiologists—Italy—Biography.
Classification: LCC QR31.L84 S45 2022 (print) | LCC QR31.L84 (ebook) |
    DDC 579.092 [B]—dc23/eng/20220228
LC record available at https://lccn.loc.gov/2022000763
LC ebook record available at https://lccn.loc.gov/2022000764

10 9 8 7 6 5 4 3 2 1

For my parents and my daughters

CONTENTS

OCTOBER 1969

Wednesday, October 15, 1969, was a beautiful crisp fall day in Boston. Across the country, American citizens took the day off from their regular activities and participated in the Vietnam Moratorium to advocate for an end to American military involvement in Vietnam. Among them was Salvador Luria, a middle-aged Italian-born biologist on the faculty of the Massachusetts Institute of Technology (MIT). Luria briefly joined the 100,000 people rallying on Boston Common before going on to meet with legislators, attend a convocation at MIT, and finally to hear Governor Francis Sargent speak at an evening peace rally in his adopted hometown of Lexington, cradle of the American Revolution.[1] As a founder of the Boston Area Faculty Group on Public Issues, Luria was a familiar figure in local events protesting the Vietnam War.

On Thursday, October 16, 1969, Luria woke up, made breakfast, and cleaned up the kitchen. A feminist at heart, he shared domestic responsibilities with his wife, Zella, a psychologist at Tufts University. As he was putting away the dishes, the telephone rang. A neighbor was calling about an exciting news item she had just heard on the radio.[2] Salvador Luria was awarded a share of the Nobel Prize in Physiology or Medicine, along with his two friends and collaborators Max Delbrück and Alfred Hershey, for his research in the 1940s on bacteriophage, the viruses that attack bacteria. The Nobel Committee recognized how their fundamental work on the nature of bacteriophage laid the groundwork for the emergence of bacterial genetics, virology, and molecular biology. By the time Luria arrived at his office in Cambridge, a telegram from the Karolinska Institute in Stockholm was waiting, confirming the good news. With a peace pin on the lapel of his sports jacket, Luria announced at a news conference that he would contribute part of his share of the prize money to the anti–Vietnam War effort. He then excused himself to go teach his regularly scheduled microbiology class before his colleagues and students descended on his office for an enthusiastic celebration.[3] The weekend passed in a blur of congratulatory telephone calls, telegrams, and letters from friends, students, colleagues, and elected officials from all over the world.

Monday morning, October 20, 1969, Luria awoke to find his photograph on the front page of the *New York Times*, with the headline, "Second H.E.W. Blacklist Includes Nobel Laureate."[4] Luria was among forty-eight scientists who had deliberately never been appointed to federal review panels for the National Institutes of Health, a branch of the Department of Health, Education, and Welfare. The *Times* reported that Luria and the other scientists still had federal support for their own research, but that for reasons of "security" and "suitability," they were barred from an advisory role in the process of science funding. Although Luria later told an interviewer that being on the blacklist was "delightful . . . [because] I didn't waste time going to Washington," in the tumultuous climate of the late 1960s, the existence of the blacklist, even one without significant consequences, served as a strong reminder to scientists that their cultural and political status was dependent on much more than the quality of their research results.[5]

On that same day, the latest issue of the liberal magazine *The Nation* hit the newsstands, with an essay by Salvador Luria entitled "Modern Biology: A Terrifying Power." In it, Luria outlined recent advances in molecular biology for a nonscientific audience, and then called for politicians and scientists to work together to create "a rational decision-making machinery" in order to use the new genetic information in a socially responsible and just way.[6] This was the first of Luria's many calls for scientists to accept a wider educational and advisory responsibility and for the general public to rely on science as a way to strengthen American social and political institutions. His cautious perspective on the social and political ramifications of molecular biology and genetic information was both a counterpoint to the excitement of his Nobel Prize and a further reminder of the crucial link between scientific research and its political context.

In this one weekend, all of Luria's public roles converged in a barrage of headlines. The intersection of science and politics—specifically the way that biology, scientific networks, political activism, public education, and political retribution were all intertwined—was a dominant theme throughout his life. The mutually reinforcing relationship between productive scientific practice and a democratic society set the contours of Luria's career as a scientist and his experience as an American citizen. His time in the United States, from his arrival as a Jewish refugee named Salvatore from Nazi-occupied Europe in 1940 to his death in 1991, precisely parallels the dramatic changes that took place in American science, particularly the life sciences, and American society during the Cold War. The ongoing navigation of scientific and democratic ideals in Luria's thought and practice reflected not only his personal development but also the changing American political and cultural context of science of that period.

Luria embraced all aspects of being a twentieth-century scientist, energetically designing and conducting experiments, publishing and presenting his findings while participating in peer review for others, applying for research grants, teaching undergraduates and training graduate students, publishing college textbooks as well as books and articles about biology for the general public, and advocating for science funding from public and private sources. He was committed to creating networks of collaborators, students, and colleagues that would act as a supportive community of researchers, and he recognized the importance of leveraging the political and cultural power of that community to secure funding to build more networks and communities. He often spoke forcefully to other biologists about their responsibility to society, paying particular attention to the possible abuses that could arise from the genetic technologies that emerged from their discoveries. But Luria also had a rich and productive life outside of science. He cooked, read literature and poetry, sculpted, and even tried his hand at the Great American Novel.[7] Most of all, he threw himself into grassroots political activity, and worked for civil rights, labor representation, nuclear disarmament, and American military disengagement from Vietnam to the Gulf War.

Luria's passion for science and engagement with the politics of freedom came from the same source and developed in parallel. His early instructors and mentors in Italy and France were men who were equally dedicated to rigorous scientific inquiry and committed to preserving freedom and opposing Fascism. After he arrived in the United States, Luria encountered scientific leaders who went to great lengths to support and protect European refugees like himself. He expressed his gratitude to them and to his newly adopted country by fulfilling the responsibilities of an American citizen, and for nearly fifty years he voted, wrote, and marched for causes he believed would advance freedom and opportunity.

Luria was an active and colorful participant in a series of "thoughtful and earnest debates over how to best make good on the promise of science to provide the greatest benefit to the largest number of people," but he was by no means the only scientist wrestling with these topics during the Cold War period.[8] His contributions included publishing a series of articles and books for the general public in the 1960s and 1970s. His perspective grew out of his belief in the liberal tradition that "assumed that an educated and informed citizenry was the major means for making decisions about the proper role of science in public life," including the use—or rejection—of genetic technologies.[9] Rather than dictate to the American people, Luria promoted scientific literacy and engaged his fellow citizens in honest and respectful conversations about science and politics.

This book is the first published account of Luria's American life in science, based on archival sources, his published work, interviews with his colleagues and friends, and government documents such as his FBI and Immigration and Naturalization files obtained through the Freedom of Information Act. His letters, books, and articles provide rich documentation of his public engagement with scientific and political issues that build up a portrait of his personality and relationships, but there are parts of his private life that remain opaque and difficult to document. In particular, Luria struggled with depression for most of his adult life. His son Daniel later recalled that Luria tried to shield him and Zella from his moods by throwing himself into his work, particularly when Daniel was a teenager, but Luria's depression still caused significant tension at home.[10] While his students noticed that he "suffered alternating moods and could be aggressive and irritable at some times while buoyant, even whimsical at others," there are no historical documents that describe his illness, save Luria's eloquent and honest description in his 1984 autobiography of his depression and the relief he eventually found from medication.[11]

Salvador Luria was a member of a remarkable group of scientists who reshaped the way we think about biology and ourselves in the second half of the twentieth century. He participated in conversations about "the volatile intersections of science and politics [that] were not abstract debates, but involved concrete political events" that had profound consequences for his own life as well as for American society.[12] Luria was also a member of another, no less remarkable group of citizens whose love for the United States and passion for democracy inspired them to work tirelessly for change in American society. His commitment to liberal causes was perceived as a threat in some contexts and as an asset in others, but in every instance, he was secure in his belief that he had the right and the responsibility to speak out as an American citizen. As a researcher and as an activist, Luria tried to help the scientific community and the United States live up to their highest ideals of equality, opportunity, peace, and beneficence. As this book will show, more than fifty years after he dominated the headlines in October 1969, his example can inspire his fellow scientists and citizens to confront that enduring challenge.

1 EARLY LIFE IN EUROPE

ROOTS

On August 13, 1912, Davide Luria and Esther (Sacerdote) Luria of Turin, Italy, welcomed their second son, whom they named Salvatore. The Luria family was neither wealthy nor poor. Davide owned a printing business and Esther raised the children. Their sons, Giuseppe, born in 1906, and Salvatore, nicknamed Totoi, attended local schools. Like other descendants of old Italian Jewish families, they employed non-Jewish servants in their home and maintained a typically religious but nonobservant Jewish Italian home—that is, they celebrated holidays with family and attended synagogue but were not strict about the Sabbath or keeping kosher.[1]

As an adult, Luria recalled that his parents were concerned with social mobility, and they saw intellectual pursuits as a means for social advancement. Giuseppe and Salvatore were encouraged to excel in school, and their academic achievements earned high familial praise.[2] One of Luria's childhood friends was Ugo Fano, the son of a mathematics professor at the University of Turin, who himself became an influential physicist. They remained close throughout their lives, and Luria's time with the Fano family in adolescence gave him a "longing for the academic life" as he became enamored of the idea of "an existence centered on intellectual rather than economic pursuits."[3]

Turin, located in the Piedmont region of northern Italy, had a small but well-established Jewish community. Jews had lived in Turin since 1424 and by the early twentieth century were well integrated into the city's political and cultural life.[4] Luria's ancestors moved to northern Italy in the sixteenth century, "about the time when the famous cabalist Isaac Luria's family went to Palestine."[5] Although none of the cabalist's published family trees include the Turin branch of the family, the extended Luria family is full of illustrious rabbis and religious leaders.[6]

From the time of the Jewish emancipation in the mid-nineteenth century until the 1930s, Italian Jews were "indistinguishable from their fellow-citizens, except that they went to synagogue instead of Mass" and to Italians, the practice of Judaism "was regarded as an amiable eccentricity rather than a social mistake."[7] Jews had access to all levels of education and taught in universities and medical schools throughout the country.[8] As full citizens, they were expected to serve in the Italian army, and several Jews served as generals during World War I.[9]

Nevertheless, as the authors Natalia Ginzburg and Carlo Levi have written, the Jews of Turin did not fully blend into Italian culture. Even the most assimilated were seen as different, and sometimes non-Jewish servants furtively tried to convert their employees and their children.[10] Through language and behavior, the Jews maintained a certain cultural distance as well. The writer and chemist Primo Levi immortalized the particular Hebrew-Italian vocabulary used by many Piedmontese Jews in his "Argon" essay in *The Periodic Table*, describing the private language that his relatives and peers used in everyday life.[11] Primo Levi's elderly relatives maintained this language as a badge of separateness, a "barrier against all of Christianity" to combat the "wall of suspicion, of undefined hostility and mockery" they frequently encountered.[12] Similarly, in 1984, on the one-hundredth anniversary of the synagogue in Turin, Levi wrote about the reserve that characterized Jewish relations with non-Jews, the need "to live in silence and suspicion, to listen a lot and talk a little, not to draw attention, because 'you never know.'"[13] That reserve notwithstanding, the Jews of Turin were loyal citizens who took pride in their contributions to the city and the region. As Levi described, "We have never been numerous; . . . and yet we do not feel that we exaggerate if we say that we have counted for something, and still do count for something in the life of this town."[14]

When he was fourteen, Salvatore had a religious crisis that started him on a path to atheism, as well as a conscious rejection of Zionism as a form of nationalism.[15] Although his family continued to practice Judaism, and for many years reminded him when the High Holidays came around, for the rest of his life Luria retained a minimal Jewish identity.[16] Nevertheless, Judaism provided Luria with a vocabulary and framework for expressing himself. In his autobiography, Luria recalled how once when his son was ill, he unconsciously began praying in Hebrew.[17] On the lighter side, he delighted in reminding Jewish colleagues that he only ate nonkosher meat.[18] He did not, however, reject the "self-imposed moral severity" of northern Italian Jews, and their idealistic vision of "reconciliation and mutual respect between people whom the religious antagonisms of the past had kept apart" as they "aspired toward a world at peace."[19] His future activism in the United States reflected his deep commitment to the most basic ideals of tolerance, peace, and ethics that he learned in the Italian Jewish community.

EDUCATION

Luria was barely ten years old when Benito Mussolini came to power after the March on Rome in October 1922. His political sensibilities were shaped by his education under the Fascist regime, although like many others his age, he did not engage in overtly political activities as a student, and he "liv[ed] passively in Fascist society."[20] Luria attended the Liceo Massimo d'Azeglio, "a famous [high] school in the Via Parini where the brightest young pupils from Turin's middle classes came to study."[21] There, Luria recalled how he "was always on the verge of failing natural science courses" because he was not "actively interested" in memorizing names and categories of objects.[22]

Luria left the Liceo with much more than a classical education. Many of the instructors at the Liceo were committed anti-Fascists, and they subtly taught their students about freedom and justice along with math, science, literature, and philosophy. The most famous of these, Augusto Monti, who taught philosophy and literature, had both a political and a personal influence on Luria. Monti was a member of the anti-Fascist, liberal socialist group of intellectuals and fighters known as Giustizia e Libertà (Justice and Liberty).[23] He was arrested along with other members of the Justice and Liberty cell in Turin in 1935, but he was eventually released and survived the war.[24] "Fiercely anti-Fascist, convinced of the ethical superiority of the classics and of the high moral calling of the writer," Monti was an imposing teacher and a great believer in the power of education to uplift the individual, as it had done for him.[25] His student Noberto Bobbio recalled that he was defined by a "spirit of seriousness and rigor, fed by a severe discipline in intellectual work and an indomitable moral energy."[26]

Monti and Luria met again after the war, and the pair recalled "how his eyes filled with tears when in the classroom he recited to us, without comment, poems of liberty; and how our band of fifteen-year-olds watched with some emotion this show of courage."[27] Monti advocated "moral revolt" based on humanist and socialist principles that included a "preference for direct democracy."[28] This blend of humanism, socialism, and democracy made the anti-Fascist movement a comfortable place for Italian Jews like Luria who had been exposed to these ideals from a young age.[29] In Monti's classroom, Luria framed the contours of his deep lifelong commitment to democracy and activism but did not yet find the courage to act on his youthful convictions.

After graduating from the Liceo in 1929, Luria entered the medical school at the University of Turin, not because of any passion for medicine but because of a combination of family pressure to enter a lucrative profession and "lack of alternative inclinations."[30] He continued to spend time with Fano, who excitedly explained the calculus and physics he was learning to his old friend as they walked around Turin.[31] Luria did well academically in medical school. In 1930, his performance earned him a place in the research laboratory of Giuseppe Levi, a well-known authority on histology, the study of microscopic structures of biological tissues, and the head of the Institute of Human Anatomy at the university.[32] Levi was a large, outspoken, athletic redhead, whose students and children lived in fear of his rage.[33] His daughter Natalia Ginzburg described how her father "appreciated and admired . . . Socialism, England, Zola's novels, the Rockefeller Foundation, mountains, and the guides in the Val d'Aosta" and did not tolerate stupidity or Fascist displays from his children or his students.[34]

From 1930 to 1936, Luria pursued several projects on the microscopic structure of muscle and nerve cells at the Institute of Human Anatomy, but these years were not particularly satisfying for the young doctor in training.[35] Luria was not intellectually stimulated by the research that he did in the histology lab, nor did he develop a close personal relationship with Giuseppe Levi, whom Luria felt at times favored other students because their parents were known in socialist and anti-Fascist circles.[36]

Nevertheless, Levi was a scientific and academic role model for Luria. He learned a "solid attitude of professionalism" from Levi: "how to take experiments seriously and bring them to conclusion, the importance of writing and publishing . . . when a set of data made sense." Levi was impressed by Luria's intelligence and his ability to formulate a scientific problem and plan the proper experiment to test his hypothesis, and Luria appreciated and later emulated the way Levi did not claim authorship on student articles unless he had actually done some of the work.[37]

During Luria's medical school years in the early 1930s, the anti-Fascist movement became stronger and more organized, and the governmental response followed suit. Giuseppe Levi was arrested and imprisoned for several months in 1934 after his son Mario and his student Sion Segre were caught trying to cross into Switzerland with anti-Fascist propaganda.[38] The elder Levi was proud to have been arrested and even tried to claim responsibility for his son's actions.[39] When Segre returned to the lab at the Institute of Human Anatomy, Luria commented to another student, "We were living next to a historical figure and hadn't even realized it."[40] Although they were not close friends, Luria later recalled how he was one of the few students who would sit in public with Segre after he was released from prison.

When Luria was in his third year of medical school, two new students joined Giuseppe Levi's lab: Rita Levi-Montalcini, a fellow Turinese Jew and one of the few women in her medical school class, and Renato Dulbecco, a serious young man from a small town in northern Italy who was thrilled to be admitted to the scientific "holy of holies" of Levi's lab.[41] These three students were only acquaintances during medical school, but after World War II, they reunited in the United States and became friends and colleagues. Luria helped Levi-Montalcini adjust to life in America when she came to St. Louis to work with Victor Hamburger in 1947, and Levi-Montalcini encouraged Dulbecco to come to Indiana to do a postdoc with Luria.[42] All three eventually won Nobel Prizes in Physiology or Medicine: Luria in 1969, Dulbecco in 1975, and Levi-Montalcini in 1986.

GENES AND VIRUSES

Luria graduated from medical school with highest honors in 1935, and his thesis "On the Relations between Body Size and Nerve Cells Size" was awarded the university's Lepetit Prize.[43] Before he could choose a path in medicine, however, he first had to fulfill his military obligation. Even though an influential uncle could have arranged an exemption, Luria served for eighteen months in 1936–1937 as a junior medical officer. Luria later explained his decision to serve in Mussolini's army—despite the fact that by then he was "a convinced but silent anti-Fascist"—as a combination of the desire to put off making a decision about practicing medicine and the desire to avoid using special protection to get out of the army. Luria spent his entire tour of duty in Italy: he was commissioned too late to serve in the invasion of Ethiopia, and his unit was not sent to support the Italian troops fighting in the Spanish Civil War (figure 1.1).[44]

After his military service ended, Luria did not want to continue to pursue anatomy research, and his limited experience during medical school and in the army showed him that he did not particularly enjoy patient care. His walks with Fano had piqued his interest in modern physics, and so he decided to focus on the medical specialty of radiology, which used X-rays and other techniques that had emerged from physics. In 1927, Hermann J. Muller had productively used those radiological tools to induce artificial mutations in *Drosophila* (fruit flies), and geneticists were conceptualizing reproduction in terms of biochemical and biophysical processes.[45]

FIGURE 1.1

Salvatore Luria in the Italian army. Photo courtesy of Daniel D. Luria.

Luria saw radiology as a possible bridge between medicine and biology—or more specifically, genetics, the branch of biology that dealt with the physical properties and behavior of genes. Luria's decision aroused Giuseppe Levi's famous wrath, and he "bellowed, 'Preposterous! You know nothing of genetics! You are not a biologist!'"[46]

Luria was undeterred, and he moved to Rome in the fall of 1937 and began his residency in radiology at the University of Rome, where Fano was doing his graduate work. His old friend helped Luria obtain a position as an assistant in the laboratory of Enrico Fermi, who was widely recognized as the star of the Italian physics community. By the early 1930s, Fermi's laboratory had attracted Italian and foreign visitors of the highest caliber to work with him and other prominent Italian physicists such as Franco Rasetti, Emilio Segré, Ettore Majorana, and Eduardo Amaldi. Italian students Bruno Pontecorvo and Renato Einaudi and foreign students Hans Bethe and Edward Teller were among the rising stars who spent time in Fermi's lab.[47]

Luria's tenure in Fermi's lab was a turning point in the development of his scientific identity. He later recalled that it was a "torment, although a delicious one" to struggle though undergraduate courses and rudimentary laboratory experiments in a new discipline. In addition to gaining important social contacts in the physics community, he learned how to approach science in a different way than he had in anatomy lab. One of the most important things that he learned that year was the need to ask and answer questions from a statistical, mathematical perspective.[48] He gained a new appreciation for asking only "concrete questions verifiable by direct experimentation," which allowed Fermi and his fellow physicists "to check the soundness of his work by nature, the infallible judge." According to Emilio Segré, "nothing pleased him [Fermi] more than combining his own theory with his own experiment."[49] In Giuseppe Levi's lab, Luria demonstrated a talent for designing experiments; working in Fermi's lab helped him develop the ability to interpret them as well.

The radiology residency did not prove as stimulating or as focused on physics as Luria had hoped. However, while he was working in Fermi's lab he encountered a field of science that seemed to combine his interests in physics and fundamental biological phenomena: radiation genetics. Franco Rasetti, one of the physicists in Fermi's laboratory who knew of Luria's interest in genetics, supplied him with key readings, including a set of papers by Max Delbrück, a German theoretical physicist who was interested in biology. Delbrück had trained with many famous physicists, including Niels Bohr, Otto Hahn, and Lise Meitner. Inspired by Bohr's famous 1932 "Light and Life" lecture, in which Bohr suggested ways that quantum theory could stimulate other scientific disciplines, Delbrück turned his attention to biology. In Berlin, he joined a study group in genetics that was "investigating gene structure by inducing mutations in *Drosophila* with x-rays," and soon decided to pursue genetics research full time.[50] In 1935, Delbrück, along with the Russian geneticist Nikolai Timofeéf-Ressovsky and the German biologist Karl Zimmer, published a daring proposal: by thinking about the gene as analogous to an atom, it could be possible to use quantitative methods to investigate the physical effects of radiation on genes.[51] In order to fully pursue this novel path, Delbrück went to the United States in 1937 on a Rockefeller Foundation grant to train at the California Institute of Technology (Caltech) in Pasadena, California, under Thomas Hunt Morgan, the giant of American genetics.[52]

Delbrück wanted to find a simpler organism than fruit flies on which to perform his radiation genetics experiments. When he arrived in Pasadena, he heard a report by Emory Ellis on techniques for investigating bacteriophages (also referred to as phages), which are viruses that attack bacteria and turn them into virus-producing machines through a process called lysis. Soon after the bacteriophage invades the bacterial cell, the bacterium bursts, releasing a new generation of infectious viruses. Bacteriophages could be placed on a petri dish or plate seeded with bacteria, and "make themselves known by the bacteria they destroy, as a small boy announces his presence when a piece of cake disappears."[53] Delbrück later recalled his delight upon discovering the ease of counting the number of viruses that had "eaten a macroscopic one-millimeter hole in the lawn. You could hold up the plate and count the plaques [of destroyed bacteria]. This seemed to me just beyond my wildest dreams of doing simple experiments on something like atoms in biology."[54]

Bacteriophages seemed like the ideal "gadget" that could be "studied operationally" and allow Delbrück to "extract numerical data, which could then be processed with mathematical analysis to construct new theories" about genetics.[55] In the early 1920s, Muller had proposed that his fellow geneticists investigate the possibility that bacteriophages could be "fundamentally like our chromosome genes," and he challenged them to embrace every method of investigating the physical nature of genes, saying, "Must we geneticists become bacteriologists, physiological chemists, and physicists, simultaneously with being zoologists and botanists? Let us hope so."[56] Although there was still considerable scientific debate in the late 1930s over whether viruses had genetic material akin to that seen in higher organisms, the prospect of finding a simple organism that consisted almost entirely of Muller's "naked genes" on which to experiment—and possibly understand the nature of that genetic material—was tantalizing for Delbrück.[57]

Shortly after reading Delbrück's thought-provoking articles in the winter of 1938, Luria struck up a conversation with his acquaintance bacteriologist Geo Rita while they were stuck on a broken trolley car. They talked "of bacteria and genes and radiation," and Rita mentioned that he was testing the waters of the Tiber for the presence of dysentery using bacteriophages. Luria was fascinated by Rita's description of his research, and when the power returned to the tram line, the two went on to Rita's lab for a demonstration. There, as Luria later described it, "Between bacteriophage and myself it was love at first sight."[58] This visceral reaction to bacteriophage animated Luria's approach to research and finding collaborators. He was delighted to have found an experimental organism on which to test ideas about genes and reproduction he had developed after reading Delbrück's articles and was further excited when he heard that Delbrück had also recently discovered the virus as an experimental organism.[59] Thus, Luria wrote in his autobiography, "As in a troubadour romance, the love triangle among Delbrück, bacteriophage, and me had come into being at four thousand miles' distance and without mutual awareness of each other."[60]

BACTERIOPHAGE

The organism that so captivated Luria and Delbrück was a relative newcomer to the laboratory setting, having been identified only twenty years earlier, and they came to it at a key juncture in its history and in the history of virus research. Although several European and American scientists were working with bacteriophages, they were more interested in the cellular, medical, or epidemiological aspects of the bacteriophage's impact on its bacterial hosts, rather than its use as a research tool to elucidate the life cycles of viruses.

Viruses were identified as organisms in the late nineteenth century, as a result of research into disease-causing microbes. Bacteria were generally isolated by passing a solution through filters fine enough to capture the bacteria but allow other, nonorganic toxins through. In time, researchers noticed that some pathological agents remained in solutions that had been filtered of even the smallest bacteria. These unknown agents were termed "ultramicroscopic filterable viruses," which was eventually shortened to viruses. Viruses were most easily identified by the ability of the filtrate to infect a host organism, and early virus research was dominated by studies of animal and plant disease agents. In the early years of the twentieth century, virus research was mainly descriptive and was dominated by projects identifying and classifying infectious viruses.[61]

In time, researchers found that bacteria themselves were vulnerable to infection by these filterable entities that were invisible under the microscope.[62] In 1915, Frederick Twort published a note in the medical journal *The Lancet* describing a curious phenomenon he had observed while working as a bacteriologist at the Brown Animal Sanatory Institution in London. He was trying to find an appropriate medium in which to grow viruses without cells or tissues as part of a larger program of research on animal diseases. One experiment with *vaccinia* (cowpox) virus on an agar medium had an unexpected outcome. *Micrococcus* bacteria that often contaminated such experiments appeared, but Twort was unable to culture them in fresh media. Rather, scattered throughout the bacterial culture were "glassy and transparent" areas in which the bacteria could not grow.[63] He "concluded that the cause of the glassy transformation was an infectious, filterable agent that killed bacteria and in the process multiplied itself," but he did not draw a firm conclusion that it was a bacterial virus.[64] "It may be a tiny amoeba [or] living protoplasm . . . or an enzyme with power of growth."[65] He mentioned that whatever it was might prove to be a significant weapon against bacterial diseases such as the dysentery facing Allied troops.

However, Twort's uncertainty about what he had discovered, along with financial constraints from the institution, prevented him from pursuing this line of research, and his work was virtually ignored for several years. It was only after the French Canadian Félix d'Herelle at the Institut Pasteur in Paris announced *his* discovery of a particle he named "bacteriophage" in 1917 that other researchers called attention to Twort's publication.

In 1915, d'Herelle was asked to investigate an outbreak of hemorrhagic dysentery among French troops stationed outside of Paris. He discovered that there was something in the stool of the recovering patients that, when added to a culture of the *Shiga* dysentery bacillus, "provokes the arrest of the culture, the death of the bacillus [and] then their lysis." In his report from September 1917 to the Académie des Sciences, d'Herelle concluded that he had found a microbe that is "an obligatory bacteriophage"—that is, a virus that literally consumes bacteria, a finding that could have significant consequences in the fight against infectious disease.[66] In this one paper, he felt that he had established "the biological nature of bacteriophage, its mode of action and its interactions with the bacterial cell."[67]

With the support of the Institut Pasteur and in several other laboratories around the world, d'Herelle performed many experiments with his new microbe on a range of animal and human diseases, and he and his collaborators added the bacteriophage to their potential arsenal of vaccines and antibacterial agents. By 1920, several of his French colleagues were working with bacteriophages, mostly at the Institut Pasteur, and by 1925, there were more than 150 publications per year on these viruses.[68]

Not everyone agreed with d'Herelle's contention, however, and some challenged his claim that bacteriophages were separate, living entities that were neither produced by the bacteria themselves nor artifacts of experimental techniques. In particular, Twort and his supporters (which included the Belgians Jules Bordet, a Nobel Prize–winning immunologist, and André Gratia) pointed out that not all bacteria were susceptible to bacteriophages, which lent support to the idea that the virus could be an enzyme produced by the bacteria itself.[69] This argument resonated with debates within the larger community of virus researchers about how to proceed with investigations of these strange organisms—if viruses really were living things at all. The resolution of all of these debates depended on the application of new techniques and new ideas from biochemistry and genetics.

For those who believed that viruses *were* living organisms, viral genetics explained the variation in pathogenicity and opened up new avenues for defining life in genetic terms. Although they behaved like "genetic entities," in the sense that they reproduced and had known stable mutations, it was difficult to perform Mendelian genetic experiments with viruses the way that Thomas Hunt Morgan and his fly researchers performed statistical genetic crosses with *Drosophila*.[70] Researchers were limited in that they could not directly observe the bacteriophage on its own; rather, they had to culture the phage in the presence of bacteria in order to detect its activity. Thus, research on bacterial and viral genetics was closely linked, and scientists working with bacteriophages also published on bacteria.

During the early years of the twentieth century, viruses were identified as organisms that were filterable, too small to be seen through a light microscope, and pathogenic to a range of hosts, and it was noted that they could not be cultured without the presence of those hosts.[71] This broad, mostly negative, definition made virus researchers vulnerable to questions about the utility of this way of identifying and describing viruses. The use of known filters showed researchers that viruses varied in size from 3 to 12 μ, so size proved to be an ineffective way to classify these organisms.[72] Furthermore, if viruses could not be cultured on their own, maybe they were not really alive.

In the late 1920s and early 1930s, new technologies developed for physics and chemistry such as electron microscopy, X-ray diffraction, centrifugation, and crystallography were applied to biological questions through research on the tobacco mosaic virus (TMV) and bacteriophage.[73] French physicist Fernand Holweck proposed "applying techniques of irradiation to determine the size of biological objects or structures, on the assumption that the targets calculated from experiments with irradiation in general corresponded to biological structures or functional units possibly too small to be measured in other ways."[74] By 1937, German scientists had perfected centrifuge and filtration techniques to the point where they were able to isolate enough pure bacteriophage for biochemical analysis, which showed that bacteriophage was made up of "approximately equal amounts of protein and DNA."[75] It was likely that bacteriophage would be the genetic tool that Muller and Delbrück hoped it would be.

By the time Luria and Delbrück encountered bacteriophage and its hosts in the late 1930s, the virus was on the brink of becoming an object of intense research as well as a potent tool for investigating fundamental biological questions. Inspired by the reports about Delbrück's research, Luria designed a set of experiments that would bombard a known number of bacteriophages (as determined by the number of bacteria they had killed) with different sized X-rays and alpha rays. By measuring the number of phage particles that were damaged by the radiation, he theorized that he would be able to calculate the size of each individual virus based on the size of the rays that had inflicted the damage. If the viruses were indeed naked genes, then this experiment would also shed light on some of the physical characteristics of genetic material. However, before he could start irradiating bacteriophage cultures, Luria wanted statistical evidence to support the assumption that a single bacteriophage killed a single bacterium. He and Rita happily worked for several months in 1938 "in the rather sleepy Institute of Microbiology" in Rome to establish the experimental basis for his future radiation experiments.[76]

CHOOSING SCIENCE

In 1938, Luria applied for and won a government fellowship to study abroad. One of his options was to go to Berkeley, California, to study biochemistry with John Lawrence.[77] Berkeley had the added attraction of being in closer proximity to Delbrück at Caltech. However, political circumstances in Italy prevented Luria from pursuing this opportunity. The news of his award came on July 13, 1938, the day before the release of the "Manifesto of Race," the document that marked the point at which "Mussolini at last . . . openly took up a position against the Italian Jews." This pseudoscientific document proclaimed that the Italian race was a pure Aryan one, and that Jews, despite having lived in Italy for close to two thousand years, "do not belong to the Italian race and are therefore unassimilable."[78] A few weeks later, Jews were banned from Italian schools. In November 1938, the "Race Laws" dictated that the Italian government would not employ Jews at all.[79] This stipulation included educational grants, and Luria lost his funding to go to America. As a Jew, he was ineligible to work in Italy as either a physician or a scientific researcher.[80]

At this point, Luria was faced with a difficult decision. He could stay in Italy with no prospects for a career in science or medicine and join the Resistance, or he could leave Italy and go to Paris and take up a temporary position at the Institut du Radium. In essence, this was a choice between political action and a scientific career. As a single young man with the means to leave a potentially dangerous situation, and with his parents' blessing, Luria chose science.[81] Armed with a letter of introduction from Fermi written in both English and Italian that attested to his "remarkable knowledge in the field of physics, his understanding of the presently most interesting problems of biophysics and his skill in devising and carrying out experimental work," Luria moved to Paris.[82]

Paris was an attractive destination for a young scientist interested in viruses, since d'Herelle was still actively researching bacteriophage at the Institut Pasteur. At the Institut du Radium, Luria met Fernand Holweck, the physicist who, in 1930, had proposed methods for using radiation biology to measure the size of microorganisms. Holweck quickly arranged for Luria to work in his laboratory through a fellowship from the National Research Fund. Luria also soon came into contact with Eugène Wollman, the chief bacteriophage researcher at the Institut Pasteur. Both Holweck and Wollman were intrigued by Luria's interest in using radiation to determine the size of bacteriophages by using X-rays and alpha rays. Wollman and his wife Elisabeth had devised several techniques for genetic research on bacteriophage, and they trained Luria in their methods for culturing the organism as a vital step in preparing his radiation experiments.[83]

Luria was productive during his time in France as he gained confidence with bacteriophage and radiation techniques. His experiments culminated in a short, densely statistical article published in *Nature* with Wollman and Holweck, "reporting the conclusion that the size of the ration target of phage was nearly the same size as the (estimated) size of the phage itself."[84] The language Luria and his coauthors used in this article underscores their focus on the physical and chemical characteristics of viruses as opposed to their genetic or biological features. Bacteriophage is referred to as a "particle," and viruses are characterized as "monomolecular structures."[85]

Despite his increasing scientific confidence and his relief about leaving Italy, Luria's social life in France was "dull," and he found Paris to be "a tense city" in a "tired country."[86] Nevertheless, he took the opportunity to explore his political identity in a free society, and he spent most of his free time talking politics and science with other Italian refugee scientists, including Bruno Pontecorvo, whom he had met in Fermi's lab. In his autobiography, Luria dated his "serious concern for politics" to his time in Paris.[87] After reading Croce, Marx, Engels, and Lenin, Luria decided that he identified not as a communist but rather as an "unstructured radical" with a deep "personal and emotional" commitment to socialist economic ideals.[88]

When World War II began in September 1939, Luria was once again in a vulnerable position. Some of his colleagues tried to escape the war by moving to the United States. Many sought the assistance of the Rockefeller Foundation, which in 1933 had initiated a program to rescue refugee scholars, particularly Jews who were prevented from working in Europe, by providing them with money and a travel visa, and helping them find academic positions in the United States.[89] In March 1940, Holweck contacted Henry Miller, the Rockefeller representative in Paris, to see if Luria could be considered for the program.[90] On April 18, Miller regretfully informed Holweck that it was "not feasible to grant Dr. Luria a fellowship at the present time."[91]

Luria was determined to get to the United States. When the Nazis invaded France in May 1940, Luria immediately applied for a visa to the United States under the Italian quota. As a Jew, he could neither safely return to Italy, where his parents and brother were preparing to go into hiding, nor stay in France, which was rapidly falling under Nazi control. As the Germans approached Paris, he fled by bicycle on June 12.[92] The night before he left, Wollman invited Luria to his home and had him listen to Beethoven "to preserve my faith in humanity."[93] Luria learned after the war that Wollman, along with thousands of other French Jews, was murdered in Auschwitz in 1943.

It took Luria a month under the hot summer sun to travel the 500 miles from Paris to Marseilles by bicycle and train. He was shot at from planes and hid for a few days in an abandoned farmhouse in the French countryside with the Italian writer Carlo Levi, all while relying on the hospitality of sympathetic strangers. When he finally arrived in Marseilles, Luria joined "a motley crew of European scholars and anti-Fascists and Jewish refugees" seeking "a French exit visa, Spanish transit visa, and Portuguese transit visa" in anticipation of receiving permission and a ticket to travel from Portugal to the United States.[94] While he waited, he took advantage of the relative quiet of Marseilles to study physical chemistry before traveling by train across Spain to Lisbon.[95]

In August, with documentation from Portuguese officials that he was a "bona- fide Jewish refugee fleeing from racial oppression," Luria was granted his U.S. immigration papers and booked passage on the Greek ship *Nea Hellas* with 800 other refugees, including the composer Paul Hindemith.[96] He arrived in New York City on September 12, 1940, with $52 and one suit.[97]

2 BECOMING AN AMERICAN BIOLOGIST

NEW YORK

In New York, Luria found American society and American science on the brink of change. Although the United States would not officially enter World War II for another year, the American government was organizing manpower and scientific expertise to prepare for the conflict, and private foundations were eager to help promising scientists from Europe find permanent positions in American academia. Luria immediately felt right at home in Manhattan and was enjoyed navigating his new life, including "the glory of the American supermarket."[1] The combination of his prior connections to elite physicists, his curiosity, wit, and good nature, and his experimental skill with viruses and bacteria together smoothed a path for Luria to integrate easily into his new surroundings.[2]

Literally and figuratively, Luria quickly shed his old identity of an Italian physician struggling to apply physics to genetic questions. Luria began the process of becoming an American citizen in November 1940, as he had no intention of returning to Italy. Despite the stress of learning a new culture and communicating in an unfamiliar language, Luria retained his sense of humor, and entertained his colleagues with the story of how an immigration officer thought he had arrived in the United States "by sheep."[3] While he was filling out his immigration papers, including the Alien Registration Form and registration for the Selective Service, he spontaneously decided to change the spelling of his name from the Italian "Salvatore" (which he claimed he never liked) to the Spanish "Salvador," and to take the final "E" from Salvatore as his middle initial.[4] When the immigration officer insisted that "E" had to stand for something, Luria asked the person behind him in line for a name that began with an "E" and became Salvador Edward Luria.[5]

By the end of 1942, Luria was offered a permanent position as a member of the botany and bacteriology department at Indiana University, and his

scientific reputation was soaring. In just over two years, then, he transformed himself from an outsider in biology and in American society to a
key figure in an influential new community of biologists, centered around
the Cold Spring Harbor Biological Laboratory, that facilitated the emergence
of several new biological disciplines, including bacterial genetics, virology,
and molecular biology. Bacteriophage was transformed too, from its earlier
use as a potential diagnostic tool into a fruitful experimental system that,
along with its bacterial hosts, reconfigured the life sciences in the 1940s.

As soon as he landed in New York, Luria contacted his old supervisor
Enrico Fermi, who had relocated to Columbia University. Fermi encouraged
Luria to once again apply to the Rockefeller Foundation for assistance and
introduced him to Leslie C. Dunn in Columbia's biology department. Dunn
was not only a distinguished American geneticist but also on the executive committee of the Emergency Committee in Aid of Displaced German/
Foreign Scholars, an organization dedicated to assisting refugee European
scientists by funding academic positions at American institutions.[6] Dunn
had seen Luria's article in *Nature* and was impressed with his work. Although
the Emergency Committee did not award Luria a grant, Dunn arranged for
him to have access to laboratory space at Columbia's College of Physicians
and Surgeons (P&S) "under a complex arrangement of minifellowships"
from the Dazian Foundation for Medical Research, the Rockefeller Foundation, and the American Cancer Society's Committee on Growth.[7]

Luria collaborated with Frank Exner, a physicist in the radiology department at P&S, using X-rays to inactivate and indirectly measure bacteriophage size, just as he had begun to do in Rome and Paris. In addition to
capturing those indirect measurements, he and Exner discovered that the
medium that the bacteriophage is suspended in has an impact on the sensitivity of the bacteriophage to X-rays. They determined that an organic solution, rather than a salt solution, provided more accurate information about
how X-rays damage bacteriophages. Unfortunately for Luria, one of Peyton
Rous's graduate students at Columbia had discovered the same thing, and
Luria had to work hard to convince Rous, an eminent pathologist, that they
had not plagiarized his work.[8]

Once he was settled at Columbia, Luria wrote to Max Delbrück to introduce himself. Delbrück was teaching physics at Vanderbilt University in
Nashville and was scheduled to attend a meeting in Philadelphia, and offered
to meet Luria there on December 30, 1940.[9] They traveled together to New
York and spent New Year's Day 1941 in Luria's office at Columbia "playing
with bacteriophage." This informal experimentation was an opportunity for

Luria to teach Delbrück the plaque technique he had learned in Paris from Wollman, as well as a time for the two to plan future experiments.[10]

Luria and Delbrück shared a deceptively simple "single common goal—the desire to understand how, during the brief half-hour latent period, the simple bacteriophage particle achieves its own hundredfold self-reproduction within the bacterial host cell."[11] Answering that basic question required answering countless questions about viral and bacterial physiology, chemistry, and genetics by devising novel statistical and experimental techniques to test their theories. Luria and Delbrück enthusiastically accepted that challenge together. There was an "instant intellectual affinity" between Max and Lu (as they called each other in letters and in person), and they decided to collaborate as often as they could, given their geographic separation.[12]

Delbrück had been invited to spend the summer at the Cold Spring Harbor Biological Laboratory on Long Island, and he suggested that Luria join him there.[13] At Cold Spring Harbor, Luria was reunited with his childhood friend Ugo Fano, who had emigrated to the United States in 1939 and was collaborating with Milislav Demerec, the director of the laboratory.[14] Under the auspices of the Long Island Biological Association and Carnegie Institute of Washington, the laboratory was on the brink of becoming one of the most important independent research institutions in American biology.[15]

Over the course of the next few years, Luria and Delbrück embraced "a publication style which emphasized few but excellent papers over many trivial ones," and shared authorship of most of their papers with each other or with students; they rarely published alone.[16] In his early papers with Delbrück, Luria was referred to as the "junior author." By early 1943, however, Luria and Delbrück were on equal footing, and Luria is listed as the first author on three out of their five joint publications.

Luria and Delbrück's initial experiments together fit into Delbrück's plan to use bacteriophage as a "gadget" for genetic research.[17] In a pair of articles published in the first volume of *Archives of Biochemistry*, they focused on an unintended research result. Their interest in reaching a range of biological audiences is evident in their decision to publish in this new journal, and their explanation was consistent with Luria's earlier characterization of "phage inactivation [as] an elementary quantic process."[18]

While working together at the Cold Spring Harbor Biological Laboratory during the summer of 1941, Luria and Delbrück tried to get behind the "closed door" of the bacterial cell wall that obscured viral growth by infecting a bacterium (*E. coli*) with two different strains of bacteriophage.[19] Their hope was that one would grow faster than the other and cause the bacteria to burst while the other one was still growing. However, what they found was "a striking case of interference." Using strains they called α and γ because those were the only Greek letters on Luria's typewriter, they determined that when a bacterium is invaded by particles of both viruses, only γ is liberated, albeit at a slower rate than if it infected the bacteria on its own.[20]

By carefully comparing the rates of growth for the two bacteriophages alone and when they were interfering, Delbrück and Luria were able to draw "some conclusions concerning the mechanism of virus growth." They suggested that virus growth inside of bacteria was contingent on the presence of "one 'key enzyme' . . . [which] may be just one molecule." Although the two viruses are very different in terms of their effect on the bacterium, they share this key enzyme as a "common factor" necessary for growth. In strongly biochemical terms, Delbrück and Luria surmised that "all previous reactions tending toward synthesis of virus α are frustrated and are replaced by the reactions leading to the synthesis of virus γ."[21]

Geneticists and biochemists had debated the possible role of such "key enzymes" in genetic mechanisms since the 1910s. Although the direct link between genes, enzymes, and proteins had not yet been made, in the 1920s and 1930s scientists agreed that life was defined by proteins, and that there must be some relationship between genes and proteins.[22] Some had even argued that genes actually were autocatalytic enzymes that provided the chromosomes with the chemical resources for self-replication.[23] Delbrück and Luria hypothesized the "key enzyme" in order to explain the observed cases of interference and to provide a more detailed picture of virus growth in general.[24]

In addition these intriguing results, Luria and Delbrück also noticed the emergence of what they called "secondary cultures" of bacteria that were not destroyed by contact with bacteriophage and seemed to be resistant to it, as well as bacteria that did not burst after infection with bacteriophage and instead became lysogenic, releasing viruses but did not undergo lysis.[25] They did not have time to investigate these observations further, since they were also expected to participate in the Cold Spring Harbor Laboratory Symposium on Genes and Chromosomes. Luria later recalled how the symposium gave physicists, chemists, and geneticists a "common language" as they took a rigorous approach to "not only the order of genes but also their nature and their function."[26] The intense seminar gave Luria the opportunity to interact with leaders of the American genetics community, including laboratory director Demerec and the famous *Drosophila* geneticist Hermann J. Muller.

Luria made quite an impression on these men in the short time he spent with them, and they supported his application for a fellowship from the

John Simon Guggenheim Memorial Foundation. Demerec saw Luria as "an enthusiastic worker with plenty of ideas and with the capacity to see these ideas through." Luria's intelligence, training and personality justified "an expectation that he will develop into a good scientist . . . he is still at the beginning of his scientific career and . . . he still has to do a great deal of work before he will be appreciated."[27] Similarly, Muller felt that "there is no doubt of his being a man of considerable ability and productive capacity, and that his training renders him particularly valuable."[28] Less than a year after he arrived in the United States, Luria was well on his way to becoming a member of the scientific elite.

Luria fully immersed himself in the American scientific community and joined several scientific societies, including the American Society of Naturalists, the Genetics Society of America, and the American Association for the Advancement of Science.[29] Although many of these organizations suspended their meetings during the war, they continued to publish valuable research articles and news of the profession. The diversity of his memberships highlights not only the fragmented nature of the life sciences in the 1940s but also the flexibility of Luria's research interests.[30] Bacteriophage had the potential to be of interest to naturalists, physicians, and geneticists alike, and Luria's engagement with each of these communities and publication in their various journals are signs of the strength of his approach to biological questions.

THE PHAGE IN ALL ITS GLORY: ELECTRON MICROSCOPE STUDIES

Luria's membership in the genetics community anchored at Cold Spring Harbor gave him access to new scientific tools and collaborative opportunities. He and Frank Exner had estimated bacteriophage size using X-rays, and Luria was eager to check their estimates using an electron microscope to visualize bacteriophage particles.[31] The electron microscope was a promising new technology that had been developed in the 1930s in Germany by a team of electrical engineers and biomedical researchers. It used a beam of electrons aimed at a dehydrated specimen mounted on thin supports in a vacuum chamber to create a magnified image with a resolution that was several hundred times higher than one from even the most powerful traditional light microscope.[32] This increased precision made it an ideal tool for visualizing microorganisms such as viruses.

When Luria arrived in the United States, there were a handful of electron microscopes in the country, the most advanced of which was located at the Radio Corporation of America (RCA) headquarters in Camden, New Jersey. In 1940, the National Research Council's Committee on Biological

Applications of the Electron Microscope funded a $3,000 a year position at RCA for Thomas Anderson, a talented physical chemist who had been unable to find a permanent position during the Great Depression, to help committee members prepare and take photographs of biological speci- mens, and to determine which features were artifacts of the preparation or microscope and which were actually features of the organism.[33] The select committee was made up of virologists, bacteriologists, and geneticists from New York, New Jersey, and Pennsylvania, including Wendell Stanley, the researcher who had crystallized tobacco mosaic virus (TMV).

Luria contacted Anderson in November 1941 to apply for a spot on his schedule. His access depended not only on gaining security clearance to visit RCA headquarters (where other researchers were working on defense contracts) but also on the approval of the committee.[34] Stanley openly dis- agreed with Luria over the best indirect method for calculating virus size, and therefore labeled Luria's proposal "controversial," but Demerec, the director of the Cold Spring Harbor Laboratory, successfully overrode Stan- ley's objections.[35]

The security clearance was approved, and Luria arrived in Camden on December 8, 1941, with batches of bacteriophage and bacterial stock. How- ever, he and Anderson soon found that his stock did not contain enough virus to be visible. Luria was compelled to culture more bacteriophage and bacteria, a task he took on as he registered as an enemy alien when the United States officially entered World War II. When Luria returned for the week of March 2, 1942, they were able to see bacteriophage "as tadpole-shaped particles, whose heads ranged from 600 to 800 Å in diameter."[36] Luria was thrilled by the opportunity to see his experimental organism directly for the first time, and he later recalled how he became "all excited" by the "beautiful architecture" of bacteriophage, embracing an aesthetic element of biological research. He had a visceral reaction to seeing bacteriophage, and he savored "the plea- sure of seeing my favorite microorganisms in all their naked beauty," each one with "a head and a tail, and whiskers, and legs like a spider's."[37]

Luria and Anderson quickly published their results in the April 15, 1942, issue of *Proceedings of the National Academy of Sciences* (figure 2.1). Their short article, accompanied by several plates of photographs of viruses and bacteria, presented a morphological analysis of bacteriophage, identifying "many extremely interesting features" of the organism. They found that phage cultures contain "particles of extremely constant and characteristic aspect," with a round head and a thinner tail that measured 80 mμ around and 130 mμ long respectively. This size was "in fair agreement" with the measurements Luria and Exner had earlier taken with their X-ray method.

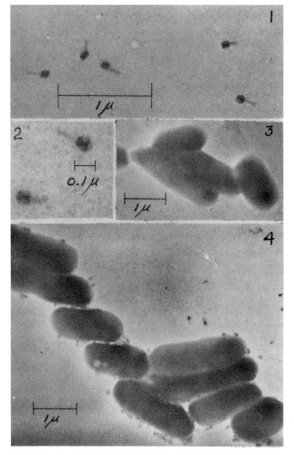

PLATE I

FIGURE 2.1

Plate I from S. E. Luria and Thomas F. Anderson, "The Identification and Characterization of Bacteriophages with the Electron Microscope," *Proceedings of the National Academy of Sciences* 28 (1942): 127–130. The large shapes in images 3 and 4 are bacteria.

In addition, Luria and Anderson added bacteriophage to a culture of *E. coli* and photographed the lytic effects of the virus on the bacteria after ten minutes. The visual evidence they collected was consistent with what Luria and his colleagues had observed through indirect experiments, so Luria and Anderson concluded that what they had photographed were indeed "actual bacteriophage particles." In addition, they cautiously noted that their evidence about bacteriophage size also provided evidence to support the "hit-theory" technique for determining the size of genes that had so enthralled Luria back in Rome.[38]

Luria was awarded a Guggenheim Fellowship in March 1942, and he planned to join Delbrück at Vanderbilt University in Nashville after they spent the summer at Cold Spring Harbor to complete a series of experiments on "phage resistance and lysogenic cultures" in bacteria.[39] Around the same time, RCA moved Anderson and the electron microscope to the Marine Biological Laboratory in Woods Hole, Massachusetts, and Luria invited Delbrück to join him there in late summer 1942. The report of their collaboration focused on the photographs taken mainly by the biologist Nina Zworykin, the daughter of Vladimir Zworykin, RCA's head of electronics research.[40] This longer paper, published in the *Journal of Bacteriology*, provided them with the opportunity to make stronger, more confident claims about the evidence from the images.

Luria, Anderson, and Delbrück focused on strains α and γ, and its *E.coli* host. They confirmed that the phage particles exhibited strain specific traits, and "by singular chance . . . obtained from a series of unusually favorable fields, pictures of a number of bacteria caught in various phases of disintegration."[41] They saw that only one particle entered each bacterium, and that there were no other particles of the same size as the virus inside the bacterial cell. This proved that "there can hardly be any doubt that the sperm-shaped particles on suspensions of bacterial viruses are the particles of virus."[42] This set of electron micrographs also demonstrated that Luria's methods for determining virus size using radiation bombardment were indeed more accurate than Stanley's reliance on filtration.[43]

In a point that they considered to be "perhaps of the greatest consequence," Luria, Anderson, and Delbrück tried to answer a basic question about virus growth inside the bacterium. In order to explain the process of infection, they drew an analogy between the viruses entering the bacterial cell and a sperm cell entering an egg. They described how once one virus infects a bacterium (or a sperm enters an egg), it somehow "bars the entry of other virus particles" of the same or different types. Although this explanation refuted Luria and Delbrück's "key enzyme" hypothesis in their 1942 article on interference, it "would lend support to those theories of the systematic position of viruses which consider it as related to the host rather than as a parasite."[44] As Delbrück explained to Anderson in a letter, "The cell is considered as a prey to two invaders, but happens to be so constructed that only one of them can succeed. The cell does not *do* anything to bring about interference. . . . The quick and specific reaction of the cell suggests that the cell is not unprepared for the attack."[45]

The electron microscope also offered Luria, Delbrück, and Anderson an opportunity to analyze the biochemical composition of the observed particles, which led to a discussion of "the tendency to speak of viruses as molecules." They traced this tendency to Stanley's crystallization of the TMV, and noted that it seemed to them "only natural that [further] chemical and physical studies have led scientists to the habit of thinking of viruses in terms of molecules." They cautioned their readers to use the term "molecule" carefully, since when describing viruses, it would mean some kind of indefinite "long chain compound." This term would not help resolve the issue of whether or not viruses are living, since according to that definition, "such 'molecules' will share with living things the impossibility of delimiting unambiguously which atoms belong to them and which do not." For this reason, Luria, Delbrück, and Anderson were hesitant to draw chemical conclusions about "the biological status of the viruses."[46] It was not until the early 1950s that the biological status of viruses was resolved through a combination of further physical and chemical studies, along with a concurrent modification, in genetic terms, of the definition of "living," based in part on genetics experiments done with bacteriophage.[47]

Luria's satisfaction with his scientific success was tempered by sad news trickling out of France. In February 1942, a "free French scientist" anonymously published a short note in *Nature*, announcing the death of Fernand Holweck under unknown circumstances.[48] Several months later, Luria received word that Holweck had been murdered by the Gestapo on December 14, 1941. He contacted Salomon Rosenblum, another refugee scientist who had worked with Holweck at the Institut du Radium, and in October, the two published an emotional obituary for Holweck in *Science*. They described Holweck's scientific contributions and praised his skill as an experimenter, generosity as a mentor, and independent character. Luria and Rosenblum lamented the circumstances of Holweck's death, noting, "It is easy to imagine that such a man would refuse not only collaboration, but even obedience to the iniquitous Nazi rule in France. He has paid with his life for his love for freedom and for his country. His example will inspire all scientists of the world in their fight for the cause of liberty and democracy."[49]

Like Giuseppe Levi had in Turin, Holweck influenced Luria as a model of the combination of scientific and political integrity. As he was busy establishing himself as an American researcher, Luria was constantly reminded of the oppressive circumstances he had left behind in Europe, and he was determined to follow in the footsteps of his teachers and mentors.

THE FLUCTUATION TEST

With Guggenheim funding, Luria moved to Nashville to collaborate full time with Delbrück in the fall of 1942, but their work was cut short by a job offer. The terms of his initial Rockefeller Foundation grant required Luria to accept the first reasonable offer he received, and late in 1942 Luria applied for and was offered an open position in the botany and bacteriology department at Indiana University.[50] In January 1943, Luria moved to Bloomington, Indiana, to begin his career in American academia. Although he had not taught before, Luria began teaching the introductory bacteriology course as soon as he arrived in Bloomington.

A few weeks later, Luria got an idea for an experiment while watching a colleague play a slot machine at a Saturday night faculty dance. Since he and Delbrück had first observed bacterial resistance to bacteriophage in the summer of 1941, he had pondered the problem of how that resistance arose. Luria considered two possible hypotheses: that the resistance was somehow triggered by the presence of bacteriophage, or that some of the bacteria were already resistant as a result of random genetic mutations. At the time, it was difficult to test these possibilities experimentally. As Luria teased his friend about gambling, he realized that mutations in bacteria could be considered analogous to true jackpots in an unprogrammed slot machine, returning small amounts infrequently, and large payouts rarely. While those returns may seem random, they can be described mathematically as a Poisson distribution, a well-known statistical model for evaluating a series of rare independent events in a given amount of time.

Following this logic, Luria hypothesized that if bacterial resistance to bacteriophage was the result of spontaneous random mutations, resistant bacteria would appear in a random distribution of "jackpots" in a series of cultures. If, however, the bacteria acquired resistance as the result of contact with the phage, resistant colonies would be evenly distributed across all cultures. In either case, the average *number* of resistant colonies across all cultures could be the same, but the *distribution* would indicate whether the cause was spontaneous or not.[51]

The next morning, Luria ran the experiment with α, a virulent strain of phage, and its bacterial host. First, he set up twenty test tubes with small amounts of bacteria, all from the same culture, as well as a larger test tube of bacteria from the same culture. Although they all started from the same stock, Luria surmised that as they reproduced over several generations, some bacteria would mutate and some would not, effectively evolving into different populations in the smaller test tubes over a matter of hours. On Monday morning, he spread bacteria from each of the smaller cultures on a petri dish already seeded with a layer of α, making sure that "all bacteria were surrounded by large numbers of virus particles," and created a parallel set of ten plates infected with α using bacteria from the larger batch as a control group.[52]

After an anxious day of waiting, on Tuesday, as he expected, Luria found several "jackpots" of resistant bacteria randomly distributed among the plates from the smaller cultures, and similar numbers of randomly distributed resistant colonies in the control group, but no "jackpots." Luria was confident that this experiment demonstrated that resistance to bacteriophage had arisen spontaneously at some point as the bacteria reproduced over several generations, rather than in response to the presence of bacteriophage.[53] Luria was fortunate to have chosen a virulent phage; if he had used a temperate strain, he would have seen that the presence of phage did indeed induce resistance and would have concluded that it was acquired rather than hereditary.[54]

Excited by his results, Luria needed someone to make sure that "there was no catch in . . . [his] reasoning," and he wanted to consult with someone who would grasp its subtlety and check his statistical thinking. "Delbrück was the obvious resource."[55] In an essay in a publication in honor of Delbrück's sixtieth birthday, Luria reconstructed their correspondence, noting that Delbrück's initial wry response was to "exhort me to go to church."[56] Within a few days, Delbrück sent him a longer answer that showed the power of Luria's novel experiment. Delbrück realized that it would be possible to calculate mutation rates by using the series of parallel cultures "to estimate rather precisely the probability that a given mutation occurred in the brief life of a bacterium."[57]

Energized and inspired by Delbrück's enthusiasm, Luria repeated the experiments a number of times to refine the results and test Delbrück's theory about mutation rates. The first footnote of their publication of this experiment states "Theory by M. D., experiments by S. E. L."[58] Their cross-country division of labor continued the pattern they had begun a few years earlier at Cold Spring Harbor, immortalized in an often-reproduced photo of them from 1941 (figure 2.2).

Delbrück was not the only one Luria consulted about his elegant experiment. Luria's Indiana colleague Tracy Sonneborn saw multiple drafts of the Luria-Delbrück paper, and he pushed the authors to consider all of the theoretical ramifications of their findings, especially in the light of his own work with paramecium uncovering evidence of cytoplasmic (as opposed to nuclear) inheritance. Sonneborn made sure Luria and Delbrück considered the broader evolutionary ramifications of their experimental proof of

FIGURE 2.2
Max Delbrück and Salvador Luria at Cold Spring Harbor, 1941. Photo courtesy of
the Cold Spring Harbor Laboratory Archives.

spontaneous mutations.[59] When Luria went to Princeton for the summer of
1943, he wrote Sonneborn, "In spare time, I am reading more about evolution
and natural selection, and improving my very weak biological background."[60]

The two corresponded regularly over the summer about the fluctuation
test. Sonneborn told Luria "the experimental work was completely unob-
jectionable and excellent," but early versions of the theoretical discussion
seemed "unsound." Although Luria and Delbrück understood their results
to be evidence of spontaneous mutations, Sonneborn was not entirely

convinced. "In the first place, I do not agree that there are only two possible interpretations to be distinguished. Second, I think that the mathematical relations would hold equally well for other interpretations than the mutation one. All that is shown is that the changes in the bacteria are more often found between closely related cells than between more distantly related ones." He cautioned Luria and Delbrück to remember that the results they observed could be due to some kind of acquired resistance that would persist through generations of asexual reproduction but were not genetic changes in the organism.[61]

Luria appreciated Sonneborn's input. He immediately wrote back, "Do not think that I did not consider very carefully the objections you put forward concerning the paper on virus-resistant bacteria." Luria defended their "mutation assumption" on the argument that the observed mutation was not only observed across several generations of bacteria, but it was also observed at a constant distribution across the population. If the resistance were the result of a physiological response to the virus, then they would expect to see "the vanishing away of the clonal, temporary resistance (or predisposition to it)."[62] He argued that because they observed the physiological response in numerous generations, it was safe to assume that it was a permanent mutation. "If this physiological status were always inherited . . . I would be ready to consider it identical with our mutation assumption. It would still be a case of stable heritable change, and it is immaterial whether the physiological property of not having the sensitive receptors for virus only appears after the action of the virus."[63]

Based on Sonneborn's comments, Luria and Delbrück "introduced some changes in the initial presentation of the hypothesis which should make it more palatable to geneticists."[64] (Unfortunately, Luria did not specify what those changes were in his letter to Sonneborn.) They were concerned about that community in particular, since they published their results in *Genetics* in November 1943. The choice to publish in *Genetics*, the journal of the Genetics Society of America, rather than in a journal of bacteriology was due in part to their close association with Cold Spring Harbor and the community of geneticists there, as well as to Luria and Delbrück's desire to use bacteriophage as a model for all genetic activity.[65]

In the article, Luria and Delbrück clearly and logically set out the various hypotheses that could explain the observed bacterial resistance to viruses, including the possibility of acquired immunity due to exposure to the virus or to genetic mutations.[66] They focused their analysis on the mutation hypothesis, and considered the additional possibility that mutations might cause immunity in a single bacterium but that those mutations would not be passed on to future generations.[67] The hypothesis of resistance due to heritable mutation predicted "the proportion of resistant bacteria should increase over time in a growing culture, as new mutants constantly add to their ranks."[68] It was crucial to have sufficient time for multiple bacterial generations to emerge, and Luria and Delbrück carefully posited "a fixed chance per time unit, *if we agree to measure time in units of division cycles of bacteria.*"[69] The theory section presented equations for "the probability distributions of the number of resistant bacteria" for both hypotheses (acquired immunity or genetic mutation), as well as Delbrück's equations for determining the mutation rate. Their quantitative data supported the hypothesis that the mutations were present before the introduction of the viruses, and Luria observed the expected increase in the proportion of resistant bacteria.

Over the course of the experiments, they found that "great variations of the proportions [of resistant bacteria] were found, and results did not seem to be reproducible from day to day." Given their hypothesis about the random presence of preexisting mutations, these frustrations were not experimental flaws. Rather, Luria and Delbrück "realized that these fluctuations were not due to any uncontrolled conditions of our experiments, but that, on the contrary, large fluctuations are a necessary consequence of the mutation hypothesis and that the quantitative study of the fluctuations may serve to test the hypothesis."[70] They confidently concluded that they had obtained experimental "proof that in our case the resistance to virus is due to a heritable change of the bacterial cell which occurs independently of the action of the virus."[71]

Although Luria and Delbrück developed the theory of the fluctuation test to explain their singular case of bacterial resistance to bacteriophage, they felt that "it will be apparent that the problem is a general one, encountered in any case of mutation in uniparental populations."[72] In 1945, Luria reported experimental evidence that not only do bacteriophage undergo mutations just as their bacterial hosts do, they often mutate in tandem as an adaptive measure. Although he cautioned that he was not claiming that bacteria or viruses have genes "in the sense of discrete material units," he called what he was seeing mutations "because of their apparently spontaneous and random occurrence, of their transmission to the offspring and of their stability."[73]

Luria and Delbrück concluded their 1943 report of the fluctuation test with the "hope that this study may encourage the resumption of quantitative work on other problems of bacterial variation."[74] Their hope was realized. The publication, which has been cited approximately 3,000 times to date, was the second most cited article in *Genetics* in the 1940s. It is considered a classic paper in twentieth-century biology and is still taught and replicated in biology classrooms.[75]

The Cold Spring Harbor community was excited about the implications of the fluctuation test, and the experimental technique Luria and Delbrück outlined spread throughout the biology community over the course of the 1940s.[76] Among the first researchers to reproduce it were Luria and Delbrück's new collaborator Alfred Hershey and Milislav Demerec, who was investigating the mechanisms of bacterial resistance with Luria's old friend Ugo Fano.[77] Demerec's work appeared in *Genetics*, *Proceedings of the National Academy of Sciences*, and *Journal of Bacteriology*, expanding the audience for the fluctuation test beyond Cold Spring Harbor.

After a hiatus because of World War II, geneticists once again gathered for the Cold Spring Harbor Symposium in Quantitative Biology in 1946. Several of the articles discussed in the symposium, devoted to "Heredity and Variation in Microorganisms," were directly based on Luria and Delbrück's methods and equations.[78] There were also immediate and significant practical ramifications for finding experimental evidence of random mutations in bacteria, since the problem of antibiotic resistance was a key issue in World War II biology research.[79] Luria himself applied the fluctuation test to the study of antibiotic resistance in a project with Eugene Oakberg.[80]

Bacteriologists and geneticists outside of Luria's Cold Spring Harbor community took up the fluctuation test as the starting point for discussions of the spontaneous origin of other bacterial mutations, as well as of the genetic basis of those mutations. While the fluctuation test was not the final word on whether bacterial genes were identical to the genes of higher organisms, it did lay to rest many older arguments against bacterial variation.[81] It also provided bacteriologists with the ability to argue that bacterial genetics could shed light on "general biological problems, particularly to those of evolution and morphogenesis." In 1947, for the first time, an article on bacterial variation appeared in the *American Naturalist*, a general biology journal that was one of the main outlets for publications on genetics and its relationship to evolution.[82]

As Luria pointed out many times over the next several years, the fluctuation test dealt a blow to the neo-Lamarckian view that the viruses somehow induced heritable mutations in bacteria. In a 1947 review, Luria commented that until the emergence of bacterial genetics in the mid-1940s, bacteriology had been "one of the last strongholds of Lamarckism," because of the difficulty in providing direct evidence for the existence of both Mendelian traits and the characteristic Darwinian criteria of random change.[83]

He stressed "how simple genetic principles may offer the lead for a correct interpretation of some of the most controversial aspects of bacterial variation."[84] His discussion of the spontaneous mutability of bacteria included evidence not only from the bacterial genetics research that he and Delbrück had inspired but also a significant amount of biochemical data. By virtue of the existence of mutations in bacteria, he argued that we may be "justified . . . in assuming the existence . . . of discrete mutable determinants comparable to genes in higher organisms."[85]

The last section of the review was devoted to "selection phenomena and evolutionary considerations," and Luria was not at all hesitant to make strong evolutionary claims based on bacterial genetics. He cited his friend Theodosius Dobzhansky's *Genetics and Origin of Species* in his discussion of how bacterial variability "provides ample material for natural selection to act."[86] He was confident that bacteria's ability to mutate spontaneously "is the mechanism that, in bacteria as well as in higher organisms, brings about a variety of phenotypes on which the environment exerts its selective role."[87] In this way, he countered the long-held bacteriological claim that changes in bacterial growth were merely temporary responses to environmental changes.[88] Luria had no doubt that natural selection operated on bacteria just like on higher organisms, and he felt that the evidence from bacterial genetics supported evolutionary biology.

Despite Luria's confidence, the fluctuation test did not convince the scientists from other life science disciplines who were actively solidifying the modern evolutionary synthesis that microorganisms followed the same evolutionary path as multicellular organisms. The evolutionary synthesis of the 1940s explained how randomly occurring genetic mutations provided a mechanism for Darwin's theory of evolution by natural selection, but the evidence came primarily from research on animals and plants. The genetic status of microorganisms such as bacteria and viruses was still unclear, and the architects of the synthesis paid scant attention to Luria and Delbrück's experimental demonstration of spontaneous mutation and evolutionary change. Other researchers, including Sonneborn and Luria's mentee Joshua Lederberg, provided additional data on bacterial genetics over the next ten years that demonstrated that bacterial genetics operate like the genetics of higher organisms, further unifying the life sciences under the banner of evolutionary theory.[89]

The fluctuation test directly inspired at least one career in bacterial genetics, a career that also had a profound effect on Luria's personal life. Dobzhansky assigned his student Evelyn Witkin to report on a preprint of the Luria and Delbrück article.[90] Her interest in bacterial genetics piqued,

Witkin eagerly accepted Dobzhansky's invitation to spend the summer of 1944 at Cold Spring Harbor, where Luria became Witkin's "prime mentor."[91] Her experiments on the mutating effects of radiation on bacteria were based in no small part on the fluctuation test.[92] Witkin's husband was a psychology professor at Brooklyn College, and one of his former students, Zella Hurwitz, came to Indiana University to do graduate work in psychology in 1944. The Witkins suggested that she contact Luria and Sonneborn, the two geneticists they had met in Cold Spring Harbor. Zella visited Luria first. Their romance blossomed quickly, and once Luria convinced her father that he could earn a living as a scientist, Zella and Salva (as she called him) were married in April 1945.[93]

INSTITUTIONALIZING BACTERIOPHAGE RESEARCH

These early papers on bacteriophage replication and bacterial genetics established Luria and Delbrück as talented experimenters within the genetics community. They began to attract a group of other scientists to their work, which prompted certain institutional changes in the way that bacteriophage researchers referred to and used their chosen organism as well as in the ways they interacted with each other. As World War II ended and the American scientific landscape was reconfigured by massive governmental investment in basic research, bacteriophage rose to prominence as a fundamental research organism, thanks to the deliberate steps Luria and Delbrück took.

Early in this period, Luria and Delbrück met Alfred Hershey, a reserved bacteriologist who was working with Jacob Bronfenbrenner in St. Louis at the time. Hershey wrote to Delbrück after reading about his early work with Ellis, and Delbrück invited him to present a paper to his colleagues at Vanderbilt University in January 1943. Delbrück was impressed with Hershey, and in April 1943, he and Luria traveled together to St. Louis for what they would later label as the "first phage meeting." Delbrück later joked that they were "a strange blend of characters 'two enemy aliens and one social misfit.'"[94] When they shared the Nobel Prize in 1969, Gunther Stent remarked, "Although it would be difficult to imagine three personalities more unlike than those of Delbrück, Luria, and Hershey, they have one trait in common—total incorruptibility—and it is just this trait . . . that these three men managed to impose on an entire scientific discipline."[95] Delbrück and Hershey published several papers together, but Luria and Hershey did not publish anything jointly, although they regularly consulted each other about their work.[96] Luria considered him a close friend, and remarked how his "writings like his experiments have a spare elegance that I greatly admire and envy."[97] Their mutual respect was strong enough to overcome the occasional disagreement over priority in discoveries and set a positive and collaborative tone for the community of phage researchers.[98]

Other researchers at other institutions soon took up bacteriophage research, and Luria and Delbrück loosely coordinated their work and continued to recruit new scientists. This community of researchers has been studied extensively since the mid-1960s in historical and sociological works.[99] However, there is a group of bacteriophage workers who have not been the focus of any attention by scholars or other scientists. There seems to have been a fairly large group of women who worked in the laboratory along with Luria, Delbrück, Hershey, and others. Nearly all of Luria's papers include an acknowledgement of the "technical assistance" of a young woman, including several who went on to earn doctorates and had productive scientific careers (figure 2.3).[100] Starting with his first publication with Delbrück in 1941, Luria thanked Edna Cordts, Nina Zworykin, Rachel Arbogast, Mrs. E. Oakberg, Mrs. E. Witkin, Mrs. J. P. Headdy, Martha R. Sheek, and Doris Templin for their help. In addition, he published several influential papers in the 1950s

FIGURE 2.3
Salvador Luria and assistant, 1948. Photo courtesy of Indiana University Archives.

with Mary Human (who was then a graduate student) as coauthor. Luria did not specify what they contributed to his experiments, so it is difficult to document their voices or activities, but they deserve to come out of the footnotes and into the historical narrative of bacteriophage research.[101]

In addition to training laboratory workers, Luria and Delbrück set out to identify collaborators and recruit graduate students, both at their home institutions and at Cold Spring Harbor. There was soon a loose network of researchers primarily in the Midwest, who called themselves the "Midwest Phage, Marching and Chowder Society" and met once a month during the winter to discuss research results and plan new experiments.[102] These meetings were intense working sessions followed by equally intense socializing: in one letter to Luria, Delbrück quipped, "Hope we do not burst the Hotel des Phages. You ought to assess everybody one bottle of whiskey."[103] Luria consistently refused to join Delbrück's hiking and camping trips, but the rest of the group happily continued their scientific conversations in the great outdoors. This group also met each summer at Cold Spring Harbor, for a few solid months of collaboration. In between meetings, they corresponded regularly and shared results, data, and bacteriophage stocks through the mail.[104]

As the number of bacteriophage workers grew, so did the potential for confusion, since there was only a loose protocol for referring to bacteria and their specific viruses using a complex series of Greek and Roman letters, as well as numbers in subscripts.[105] In the summer of 1944, at Cold Spring Harbor, Delbrück sought to standardize the different strains of bacteriophage and the nomenclature used to describe them, and so he organized a "Phage Treaty." All of the workers agreed to "standardize the phage systems, as well as to establish the Phage Information Service."[106] They would concentrate on *E. coli* and its seven known phages, and named those phages T1 through T7, so that stocks could be shared, and results could be reliably compared.[107]

Delbrück and Luria needed a way "to recruit enough students to maintain the momentum created by their initial discoveries."[108] This issue weighed heavily on their minds not only because they were so excited about the possibilities of their findings but also because they were so concerned with attracting the "right kind of researcher."[109] Luria later described how their "correspondence between 1940 and 1943 dealt, more often than not, with the problem of attracting the interest of geneticists, biochemists, and cell physiologists to the dimly glimpsed green pastures of the promised land."[110] World War II also hampered their efforts: most scientists who had not been drafted were working on defense-related projects.[111]

Many years later, "Max recalled that in the fourth summer [of working together at Cold Spring Harbor], Luria suggested that they bring people into phage by giving the summer course."[112] While most courses at Cold Spring Harbor had been geared toward college students, this course was meant for scientists who were already trained in other fields. The idea was to give a short course in how to handle bacteriophage and then let people start experimenting. Delbrück organized and taught the first phage course in 1945, and for the next twenty-five years, "nearly all of the instructors . . . were either former students in the course, or had other connections to Delbrück."[113]

Delbrück set high mathematics prerequisites for the course in order to attract scientists with similar backgrounds to his own. The students spent time learning how to prepare bacteriophage and bacteria for experiments, and then reviewed the techniques that Luria and Delbrück used in their growth and mixed infection experiments. As they "retraced in nine days Delbrück's eight-year path," they learned how to combine relatively simple laboratory techniques with sophisticated mathematical and theoretical analysis.[114] Although they worked on the border between biology, physics, and chemistry, Delbrück and Luria wanted the phage workers to learn to think like physicists, just like Luria had in Fermi's lab in Rome.

In August 1945, six people met the prerequisites and enrolled in the course. Demerec's director's report from that year described the founding of the course and noted, "it was very successful."[115] As atomic bombs dropped on Hiroshima and Nagasaki and World War II ended, four men and two women learned how to apply physics to bacteriophage. The date may have been a coincidence, but it does foreshadow a broad trend in the physical and biological sciences. In the post-Hiroshima era, there was "a postatomic enthusiasm for a new physics of life."[116] Using the laws of physics to explain rather than to end life was an attractive prospect for many atomic scientists.

In the next few years, the course attracted a number of highly trained scientists from a range of fields; in 1948, Demerec pointed out "of the 47 individuals who have taken this course so far, 30 have been active research workers with a doctorate."[117] In the summer of 1950, the phage meetings that had been in other cities during the winters in the previous few years moved to Cold Spring Harbor "immediately following the end of the course, so that new recruits could be inducted straight-away into the phage community."[118] The summer phage course was taught for twenty-five years, and attendance has been a badge of honor among biologists.[119]

Luria slowly began attracting graduate and postdoctoral students to his laboratory in Bloomington. James Watson was the first to join, in 1947, and Luria assigned him a project on what was then called "multiplicity reactivation," the phenomenon in which individual phage particles that were inactivated by X-rays or ultraviolet light would somehow still produce active phage that would grow inside the bacterium and ultimately lyse it when they infected a bacterium in a group.[120] Watson chose Luria as his graduate advisor in part because he knew that Luria and Delbrück, whom he had admired from afar, were regular collaborators. Watson, who was only nineteen when he started graduate school, later described "growing up in the phage group," and recalled car trips to visit Hershey in St. Louis, émigré physicist Leo Szilard in Chicago, and Delbrück while he was in Nashville.[121]

After World War II ended, Luria regularly went back to Italy to visit his parents. On one of his trips to Turin, he visited Giuseppe Levi's lab and encountered his old colleague Renato Dulbecco.[122] Like Luria, Dulbecco did not want to practice medicine and was intrigued by the intersection of modern physics and biology. Luria invited Dulbecco to join his lab in Indiana as a research associate, and he accepted in 1947. Luria and Dulbecco tested Luria's ideas about the possible recombination of genetic material into a bacteriophage that would still be virulent, and jointly published a quantitative analysis of the phenomenon, which had tantalizing implications for viral genetics, in *Genetics* in 1949.[123]

SCIENTIFIC ACTIVISM THROUGH THE GENETICS SOCIETY OF AMERICA

Membership in the Genetics Society of America (GSA) gave Luria not only a journal in which to share his scientific findings but also a mechanism within the American scientific community to provide support to scientists throughout the world who were affected by the disruption and destruction of World War II. His colleagues in Indiana, particularly Ralph Cleland and Hermann Muller (who joined the department in 1945), were leaders in the GSA's Committee on Aid to Geneticists Abroad (CAGA). In 1946, the committee requested donations for "deserving foreign geneticists who have not helped the Fascist or Nazi causes, or who are known to have suffered by reason of their opposition to these causes," as well as for "reputable geneticists and their immediate families who through no fault of their own find themselves in displaced persons camps."[124] The donations would be used to send food and clothing through the Cooperative for American Remittances to Europe (CARE) and to help European geneticists find positions in the United States if they wished. Luria and Delbrück helped identify twenty-one geneticists, mainly in Germany, who needed assistance, and the committee spent a total of $424 on forty CARE packages. Cleland gratefully reported to the GSA that the food packages "make it possible for scientists to spend less time foraging for the means to prevent starvation, thus freeing time for research."[125] Thus, CAGA also facilitated the distribution and exchange of publications and other materials for geneticists across Europe, including in the Soviet Union.[126]

Soviet geneticists were of particular concern to the American genetics community because of the difficult ideological challenge posed by Soviet agronomist Trofim Lysenko.[127] Under Joseph Stalin, Soviet science was supposed to be oriented around meeting the needs of the people, rather than pursuing knowledge of the natural world for its own sake as it was in the capitalist West.[128] In the late 1920s, Lysenko developed a technique that he called "vernalization" that he thought would make winter wheat plantable in the spring, which could help the Soviet Union overcome food shortages and increase the efficiency of its collective farms. He applied his technique to a number of crops, and by the mid-1930s, Lysenko had a theory to explain the process that rejected the gene as the material basis of heredity and embraced the inheritance of acquired characteristics. According to his "Michurinist" biology (named after a Soviet plant breeder who died in 1935), plants go through distinct developmental stages. By changing the environmental conditions of the plant at the end of each stage, researchers could permanently alter the heredity of the plant, thereby changing its developmental pattern. Lysenko's theory stood in stark ideological and experimental contrast to Mendelian genetics, which rested on the assumption that the germ plasm is fixed and environmental factors alone do not permanently alter the heredity of an organism.

In the 1930s, despite the fact that none of his agricultural projects were successful, Lysenko was rewarded for this service to the Soviet people with election to the Lenin All-Union Academy of Agricultural Science, and was elected president in 1938. The period of his leadership coincided with the height of Stalin's Great Terror, when scientists from all disciplines—including a number of well-known classical geneticists—were among the eight million Soviet citizens who were arrested and sometimes executed on ideological or antisemitic grounds.[129] By the late 1940s, Lysenko's dominance of the Soviet genetics community emerged as a public relations problem for geneticists around the world, particularly for left-leaning scientists who approved of Soviet state control of science but were unable to support Lysenko's methods or conclusions.[130]

In 1945–1947, the leaders of the American genetics community launched a coordinated campaign to support Soviet geneticists and to discredit Lysenko on scientific grounds. One of their tactics was to publish reviews of Lysenko's book *Heredity and Variability* in major American and British scientific journals.[131] Luria's friends and colleagues Muller, Dobzhansky, and Dunn invited him to join a group of geneticists who chose to focus on the flaws in Lysenko's science since "adverse reviews of Lysenko's book by reputable scientists in Western countries would be seriously considered in the USSR and have a beneficial effect for Genetics there, at the same time weakening Lysenko."[132]

In the spring of 1946, Luria drafted a letter to J. B. S. Haldane, a prominent British biologist who was also a Communist Party member, on behalf of the campaign.[133] Luria wrote:

> It is our feeling that the initiative to explain the danger represented by Lysenko's ideas to the authorities responsible for scientific research in the Soviet Union should come from somebody who, besides being respected as [a] scientific authority, could not easily be attacked by Lysenko as a political enemy of Soviet Russia, [or of] using his prestige for the purpose of either a personal attack on Lysenko or of a campaign to defame Soviet Science. We think you are the person most likely to lead successfully such an action, which would render a great service to the Soviet Union and to world biology.[134]

With some modifications from Muller, the letter was sent out with the signatures of this elite group of American geneticists, although Luria was unsure whether to sign, "since Haldane has probably never heard my name."[135] Haldane refused to write the review and returned the original letter to Muller.[136] Haldane struggled to reconcile his political commitment to Marxism and his scientific commitment to Mendelian genetics, and he ultimately resigned from the Communist Party when he could no longer defend Lysenko's scientific work.[137]

The Lysenko affair came to a crisis point in 1948. At the Lenin All-Union Academy of Agricultural Sciences meeting in August, Lysenko announced that his brand of Michurinist biology, favoring a malleable view of heredity and rejecting chromosomal inheritance, had the full support of the Communist Party and the Soviet government. This announcement cemented his control over Soviet biology and forced many researchers to recant their earlier views or face severe punishment. Soviet biology would be reorganized around Lysenko's version of genetics, and Mendelian genetics was eliminated from curricula and research projects.[138]

The GSA struggled to formulate an appropriate response because of concerns about international and American political backlash, and the effort to articulate a pointed critique of Lysenko was bogged down by procedural disagreements.[139] In his role as GSA president, Luria's colleague Tracy Sonneborn appointed a Committee to Combat Anti-Genetics Propaganda (CCAGP) in late 1948. However, the group was not authorized to speak on behalf of the entire organization, and so its members, including Dobzhansky and Muller, published articles in the scientific and popular press denouncing

Lysenko on their own. At the GSA's annual meeting in December 1949, there was vigorous debate about whether to combine the CCAGP with the GSA's executive committee into a new Committee of Nine that would "be empowered to speak and act for the Society on matters of public concern in which the Society has a vital interest." After a motion to delay all discussion on the issue for a year did not pass, the original proposal was then amended to call on the new Committee of Nine to "carefully consider alternate proposals with regard to provisions for a Committee to speak for the Society in matters affecting freedom of science" that would present a plan to the society at the September 1950 meeting.[140]

Rather than focus solely on the challenge represented by Lysenko, the Committee of Nine proposed that the GSA form a standing Committee on Public Education and Scientific Freedom with a broad mandate to "take action on all public matters of concern to the Society," but it would speak and publish "solely *as a Committee of the Society and not in the name of the Society as a whole.*"[141] This new committee would be the voice of the GSA on many public science issues, including "the contribution of genetic knowledge to debates about atomic energy and atomic bombs, and the bearing of medical and commercial practices, such as the use of radiations, on the genetic welfare of the people." To that end, the committee should plan a "long range policy of public education in the methods, principles, and applications of genetics."[142]

Luria was one of nine geneticists nominated to join the committee. His candidate statement is a clear articulation of his early beliefs about the public role that geneticists—as individuals and in groups—should play in American life. He doubted that the committee would have any influence on the state of genetics in the Soviet Union, so rather than focus on Lysenko, he felt that it could contribute to "the education of the public in the fallacy of racial prejudice, and the defense of scientists who may be victims of attacks on academic freedom."[143] Although he did not earn enough votes to serve on the GSA committee, he soon turned his attention to those issues on a local and national level on behalf of other organizations.

AMERICAN CITIZENSHIP AND SCIENCE

After six busy years as an active scientist in the United States, it was almost a formality when Luria became an American citizen on January 10, 1947.[144] Nevertheless, Luria was excited at the prospect of his new status. On January 6, he wrote to Delbrück, saying, "On Friday I'll be an American, unless . . ." and on January 24 ended a letter with a casual, "By the way, I'm an American now."[145] Luria's colleagues and friends offered congratulations and welcome to the newest citizen.[146]

Luria participated in the American political process even before he became a naturalized citizen. When he first arrived in Bloomington in 1943, Luria became involved in the campaign of an independent candidate for Congress through the influence of his friend and colleague Kenneth Cameron. The candidate was unsuccessful, but Luria learned from the experience. He not only observed the process and procedure of American electoral politics, but he "also realized the importance of choosing the issues on which to do battle." As a member of a group of "progressive intellectuals," he saw that "electoral campaigns are not a suitable vehicle for long-term political organization of radical forces."[147]

Soon after he and Zella got married in 1945, they formed a two-person campaign for the Democratic candidate for Congress in New York's Suffolk County. The candidate, affiliated with the American Labor Party, had agreed to run only on the condition that he did not have to campaign in person. The Lurias saw this as "a splendid opportunity to slay the dragon, visit the countryside, and make new friends." They toured Long Island, talking to the laborers who had been overlooked by the moneyed Republican bosses: hospital workers, shopkeepers, servants, gardeners, and housekeepers. Their work was successful, and Luria was proud to report "since then that Congressional district has sent a Democrat to Congress eleven times."[148]

This experience inspired Luria to think about labor causes, and he joined the small but vocal Indiana University Teachers Union chapter of the American Federation of Teachers (AFT) when they returned to Bloomington in the fall.[149] The AFT did not limit its activities to on campus labor issues facing professors, instructors, and graduate students. They encouraged the citizens of Bloomington to vote in local elections and urged local government to maintain rent control despite the end of a housing shortage.[150] As president of the chapter from 1948 to 1950, Luria was particularly engaged with a 1950 effort to desegregate restaurants in downtown Bloomington, and invited the local chapter of the NAACP to a meeting of the Indiana University AFT to learn more. When a number of restaurants closed their doors rather than integrate, Luria was particularly distressed by the "flippant" way the university's student newspaper covered the closures. He wrote an irate letter to the editor, chastising the paper for not informing the student body about an important attempt to overcome discrimination and injustice against fellow students.[151]

In the first presidential election after he became a citizen, Luria exercised his voting rights for the cause of peace, supporting Henry Wallace's ill-fated run for president in 1948.[152] Luria was attracted to Wallace's vision of America as a world leader for peace and democracy, and supported the citizens' movement that seemed to be a "genuine enough new political movement against the social retrenchment of the Truman years and the beginning Cold War, with its testing of atom bombs and the growing danger of nuclear war."[153] He was an Indiana delegate to the raucous Progressive Party convention in July of that year, and served as a poll watcher on election night.

The campaign to elect Henry Wallace president of the United States in 1948 was repeatedly dogged with allegations of Communist control, and his supporters were often accused of disloyalty. On the basis of Luria's public support of the candidate and other earlier political affiliations, the FBI began actively investigating him in January 1950. FBI director J. Edgar Hoover directed the special agents in charge (SAC) in regional offices in New York, Boston, and Indianapolis "to conduct a thorough discreet investigation" of Salvatore Luria, under the designation of "security matter-C."[154] The investigation was based on "an allegation from the State Department that he is a Communist."[155] Despite this claim from the State Department, there is no evidence in the FBI file that Luria was ever a member of the Communist Party.

The Indianapolis SAC called on ten of its reliable confidential informants in the area and submitted a ten-page report on Luria on July 7, 1950.[156] This report, especially the section on Communist Party activities that lists each confidential informant's assessment of Luria's scientific credentials and political activities, shows us what Luria's contemporaries considered "Communist Party activities." None of the informants knew whether Luria was a member of the Communist Party, but all of them commented on his liberal views with regard to race relations and labor issues. Questions about Communism, it seems, were a broad invitation to comment on all types of political ideologies. The distance between a liberal stance on American domestic politics and loyalty to the Soviet Union was thought to be quite short.

The report begins with a "synopsis of facts." After summarizing Luria's employment history in the United States, the synopsis outlines the key evidence for suspecting him to be a Communist. "Subject described by informants as a very liberal thinker who has been active in labor and racial questions in Indiana. Subject was also active in the Progressive Party in this area during the last national election. [Informant's name blacked out] believes LURIA does not strictly conform to the CP [Communist Party] line but is sympathetic with most of CP's theories. Background set forth."[157]

The report goes on to give details of Luria's residence and marital status, employment history, education, and family background, all obtained from confidential informants. This section includes a physical description that lists Luria's complexion as "swarthy" (as Jews and Italians were often described) and notes that he is "always attired very neatly, usually wears business suits rather than sport clothes" (figure 2.4). Under "characteristics," the SAC wrote "has slight foreign (Italian) accent."[158]

The information in the section on Communist Party activities is a mixture of scientific commentary and statements about local issues and university politics. Their names are blacked out, but most of the informants appear to be fellow scientists and other academics from Indiana University, since most of them state that they met him through the university. No one made any direct allegations about Luria's membership in the Communist Party, but all of the informants felt comfortable commenting on his political orientation,

FIGURE 2.4
Salvador Luria in his office at Indiana University, 1947. Photo courtesy of Indiana University Archives.

even if they admitted they had never discussed politics with him. They also volunteered information on the quality of his science.

The first informant described Luria "as the most liberal-minded professor in the University." He mentioned how Luria, as a member of the Teachers Union, "once wrote the President of Indiana University demanding the reason why no Negroes were serving on the faculty of the University." The informant claims "this was merely an example of how Luria is active in racial questions, although he has never affiliated himself with other local organizations in racial disputes." Liberal attitudes aside, the informant "stated that he knew nothing specific concerning the loyalty or disloyalty of Professor LURIA to this country . . . [and he] knew of no subversive organizations to which Professor LURIA might belong."[159]

The next informant, also a faculty member at Indiana University, admitted that he had "never actually discussed political theory with LURIA," but because of Luria's interest in labor questions, "together with opinions expressed to him by other members of the faculty, lead him to believe that LURIA is very liberal minded." This informant also had no concrete information on Luria's membership in any subversive groups, "nor did he know anything derogatory concerning LURIA's loyalty to this country." Another informant stated honestly "that he had no definite evidence . . . [to confirm his suspicions about Luria's membership in the Communist Party in Europe] but it was merely his opinion based on the subject's attitude and actions." All he could say with certainty was that Luria was "a person who is very idealistic in political theories."[160]

Several of the informants mention Luria's status in the scientific community while discussing his political orientation. One "considers LURIA an excellent scientist but stated that he has rather radical ideas" which were evident when Luria "publicly complained of the treatment of Negroes in restaurants in Bloomington." Another "stated his belief that LURIA is probably the top third scientist in his field in the United States today, but in his opinion, LURIA has very liberal and socialistic ideas." Despite his commitment to socialistic ideas, this informant felt that "LURIA does not follow the Communist Party line all the way."[161]

One informant "reasoned" that if Luria were following that line, then he could not disagree with Lysenko's controversial rejection of Mendelian genetics. "However . . . although LURIA was in disagreement with this theory, he was sympathetic enough towards LYZENKO [sic] to object to his being publicly denounced." That informant claimed that Luria had participated in the attempt to delay the executive committee of the Genetics Society of America (GSA) in denouncing Lysenko and his theory in 1949. Although most of the nearly 400 members present agreed with the proposed statement, the informant described how a "small bloc" of members, including Luria, "used technicalities of parliamentary procedure to defeat this motion" and forced the establishment of a committee to study the matter. "In effect, this held up a public denunciation by the Genetics Society of the LYZENKO [sic] theory for a period of one year."[162]

At the time, geneticists dismissed any allegations of disloyalty or Communist sympathy on the part of the members of the GSA, even those who tried to delay the GSA's response. In the March 1950 issue of *Scientific American*, Carl Lindgren, a geneticist from Southern Illinois University, reported that while the proposal to have "a small group of the Society be empowered to speak for the Society 'as a whole'" was defeated, "This does not mean that there is a single member of the Society who respects or supports Lysenko; if such a member exists, he did not make his presence known."[163]

The unknown FBI informant's accusation against Luria is completely false.[164] On the contrary, a few months later, he was nominated to serve on the new GSA Committee on Public Education and Scientific Freedom. At the time, no one in the FBI felt that this accusation was significant, and it was not considered to be evidence that Luria was a Communist. The July FBI report from Indianapolis concluded with a list of leads for other regional offices to pursue, and the New York and Los Angeles offices were instructed to "conduct appropriate investigation(s)" on Luria.[165]

The Indianapolis office also instituted a mail cover, an illegal but common practice of the FBI.[166] Mail covers involved monitoring to whom the subject wrote and who wrote to him, and opening letters to and from the subject. Agents soon opened several letters from Luria so that they could analyze his handwriting and compare it "against the Bureau's file of handwriting specimens of unknown Espionage Agents."[167] They noted that Luria received several letters from addresses in Turin, presumably from his family, as well as a letter from the American Civil Liberties Union in Chicago, and two letters from "SA" with a New York address.[168] The New York office was requested to furnish information on this mysterious addressee.

A few weeks later that office sent a letter to the FBI director informing him that the "SA" at that address was the editorial office of *Scientific American* magazine. A scientist corresponding with the editors of a magazine about science should not be regarded as suspicious, but this was no ordinary science magazine.[169] In 1949 and 1950, two confidential informants, including Whittaker Chambers, "self admitted former Communist espionage agent," told FBI special agents that some of the editors at *Scientific American* were Communist sympathizers who had left *Time* magazine to work at *Scientific*

American. Through them, the magazine "might be a medium through which the Communists are obtaining classified information of a scientific nature." The editor "would be able to obtain a considerable amount of information of a classified nature during his discussions with authors of articles on the atom bomb, the hydrogen bomb, etc."[170] Indeed, the magazine had recently had some trouble with sensitive material. In April 1950, the Atomic Energy Commission ordered the destruction of the publication plates and 5,000 copies of *Scientific American* with an article by Hans Bethe arguing against the use of the hydrogen bomb, on the grounds that "certain material in it should not be published."[171] Luria did not publish anything in *Scientific American* until 1955, but the suspicion that his correspondence aroused is an indication of how anxious the FBI and the anti-Communist movement was about the loyalty and public activity of scientists. Without evidence to exonerate him, the FBI investigation continued.

THE EMERGENCE OF VIROLOGY AS AN INDEPENDENT SCIENTIFIC DISCIPLINE

Luria's political affiliations may have aroused suspicion, but his stature in the scientific community was secure. His clippings file at Indiana University contains notices about his invitations to speak at conferences in England, Italy, Sweden, and Denmark, as well as proud reports of his being awarded generous grants from the American Cancer Society to support his research.[172] In 1948, he delivered the first Slotin Memorial Lecture at the University of Chicago, dedicated in memory of a biophysicist who had been killed in an accident during atomic bomb development in Los Alamos in 1946.[173]

In 1950, Luria was invited to give the prestigious Jesup Lectures at Columbia University.[174] As L. C. Dunn described the series, it was an opportunity for "those who have made significant contributions to modern biology" to present "a general problem in such a way as to be intelligible to students of biology."[175] Luria chose to discuss bacteriophage research under the broad question of reproduction. This served not only as a way for him to link his work to the genetics legacy of Columbia University but also to tie virus research to one of the most basic questions in the life sciences—how does one organism give rise to future generations? He stated his ultimate goal plainly: "justifying the claim of my field of research to recognition as a branch of fundamental biology."[176] At the time, he noted, virology was in a "borderline position . . . between biology and protein chemistry," but Luria was ready for it to move firmly into biological territory.[177]

Over the course of seven lectures, Luria laid out nearly a decade's worth of experimental evidence on bacteriophage morphology, genetics, and behavior. Before he began, he offered an operational definition of all viruses that simultaneously accounted for their unique features and made them typical biological objects. In Luria's view, a virus is "an exogenous submicroscopic entity capable of reproduction inside living cells only."[178] As his career thus far had demonstrated, bacteriophage research also required a certain amount of bacteriological research, since the hosts are a key part of the system. The fact that viruses penetrate cells from the outside is what makes them independent biological objects, capable of evolutionary development. Similarly, the evidence that virus particles are the only ones released from an infected cell indicated that they undergo a form of biological reproduction, as opposed to chemical synthesis.[179]

While he certainly privileged the genetic approach to bacteriophage research in these lectures, Luria confidently related virus reproduction to problems in several established biological fields, including cell biology and population genetics.[180] One lecture reviewed recent advances in animal virus research, emphasizing how the emerging data on their morphology and adaptive qualities was useful not only for disease research but also for basic biological problems.[181] Luria devoted an entire lecture to biochemical studies of bacteriophage, although he cautioned "we would be in for big disappointments if we were to rely too much on biochemistry in studying the problem of reproduction" because of technological and methodological limitations inherent in chemistry.[182] True to his dedication to the larger biological picture, Luria spent the last lecture exploring possible paths in viral evolutionary history. He concluded his review of the ways in which virus research could shed light on more general biological questions about reproduction by offering viruses as a model for an even more basic problem in evolutionary history, how the cell became "the operational unit of life."[183]

Luria's wit and pleasure in his research emerged clearly in these lectures. He noted that the process of virus replication "had until recently been completely protected from indiscreet eyes by the apparent inviolability of the inner bacterial privacy."[184] In the third lecture, he introduced the genetic analysis of virus reproduction by comparing the relationship between the bacterium and its phages to a mystery novel, "in which the detective attempts to reconstruct the crime from evidence derived from all possible directions."[185]

Luria returned to these general themes again that spring in "Bacteriophage: An Essay on Virus Reproduction," one of the core readings for the "Viruses 1950" conference, held at Caltech in March 1950, and published in May 1950 in *Science* magazine.[186] He framed his discussion in the broadest possible terms with regard to virology and genetics. The article delineated the disciplinary boundaries of the emerging independent field of virology, while defining the "ultimate goal" of all biological research as "the identification of the elementary 'replicating units' of biological material and the clarification of their mode of reproduction."[187] With two different but equally important audiences—virologists on the one hand and members of the American Association for the Advancement of Science on the other—Luria's argument for the role of virus research in reaching the "ultimate goal" of biology had to be a strong one. He was confident that bacteriophage research would continue to shed light on the process of the reproduction of genetic material within all cells.[188]

Luria's ability to communicate his excitement about bacteriophage and its importance served him well as he considered how to reach yet another audience when he published *General Virology*, the first comprehensive college textbook in the field. The text was an outgrowth of Luria's teaching experience in Indiana. In the preface, he recalled how in 1946 he began planning a new course in virology, one that was not "a watered-down course in virus diseases" but rather "a new type of course, in which virology would be presented as a biological science, like botany, zoology, or general bacteriology."[189] As far as he and his colleagues knew, this new virology course was the first of its kind: a comprehensive survey of viruses taught outside of a medical context.[190] Luria acknowledged the difficulty of writing a textbook for "a science developing as fast as virology," since "original work is proceeding at such a pace that interpretations presented in research articles must often be revised in galley proof." Nevertheless, he felt that the time was right to attempt "a provisional integration."[191] That integration, which, like the Jesup lectures, reflected Luria's commitment to relating virus research to fundamental biological questions, was a founding moment for virology as an independent branch of experimental biology.[192]

LEAVING INDIANA

In the spring of 1950, Luria received a job offer from the University of Illinois at Urbana-Champaign. This was not the first time he considered leaving Indiana University, and political issues had a role in his earlier negotiations with the administration about whether he would stay. In 1947, he considered moving to a research position at Cold Spring Harbor. However, the Indiana administration was able to entice Luria to stay on as a member of their newly independent bacteriology department, despite CSHL's offer of a $6,000 salary.[193] In 1948, Iowa State College offered Luria $7,000 to join their faculty. Immediately, Ralph Cleland and Leland McClung contacted Indiana's Dean Ashton and asked him to help them keep Luria. McClung was concerned because another member of the bacteriology department, Irwin Gunsalus, had just received a competitive offer as well. If both men left, the university would be faced with the expensive prospect of replacing two first-rate scientists.[194]

Both McClung and Cleland mentioned that Luria's teaching and advising abilities had improved a tremendous amount since his arrival in 1943. Cleland urged Ashton to match Iowa's offer, because Luria was "unquestionably one of the leaders in American biology at the present time." His presence in Bloomington was a crucial part of Indiana's excellent international reputation, and if he left, "its outstanding reputation in this field . . . [would] collapse." Cleland also acknowledged Luria's role in on-campus politics. "While I do not approve of his activity in connection with the Teacher's Union, I do not think this should enter into the situation. In fact, failure to retain Luria would be interpreted by those so inclined as based on his activity in this regard, and this feeling would be damaging to faculty morale."[195] McClung acknowledged "that in previous years there was some doubt in my mind concerning the desirability of retaining Luria," but he now felt that it was important for the university that he stay.[196]

In 1949, Luria and Gunsalus each received offers from the University of Illinois at Urbana-Champaign. Watson had finished his Ph.D. and was heading to Europe to study biochemistry and Dulbecco had also completed his time in Indiana, but Luria's postdoctoral researcher Giuseppe Bertani was invited to move along with him. This time, only Herman Muller lobbied the administration on Luria's behalf.[197] The other Indiana professors still liked and admired Luria, and they wrote warm letters of recommendation to Illinois on his behalf, even though they were loathe to see him leave.[198] Sonneborn was honest about his feelings while praising Luria to H. O. Halverson, chairman of the Illinois department of bacteriology. "If I were to allow my selfish interests and my loyalty to Indiana University to dictate this letter," he wrote, "I would withhold what I know to be the facts about Professor Luria in the hope that this would result in his staying here." Sonneborn graciously congratulated Halvorson "on inducing . . . [Luria] to be willing to join your department."[199]

In March 1950, Halvorson offered Luria a salary of $8,000 a year, and the Indiana administration did not try to match the offer.[200] In a handwritten note on Muller's copy of his recommendation for Luria to the faculty at Illinois, he commented, "The Biologists wanted him retained at the faculty here but as far as I know the Trustees were against him because Luria was too left."[201] The Luria family, which by then included Daniel, who was born in 1948, moved to Champaign in the summer of 1950.

3 FREEDOM FOR SCIENCE IN COLD WAR AMERICA

"ESPIONAGE—R, ATOMIC ENERGY ACT"

In September 1950, the Springfield, Illinois, FBI office took over Luria's investigation. On October 24, 1950, Hoover sent an urgent teletype to the special agent in charge (SAC). "In view of subject's association with Bruno Pontecorvo, British atom scientist who has disappeared, apparently into Russia, according to recent news items, conduct immediate thorough investigation and expedite handling of all leads."[1]

Bruno Pontecorvo was an Italian Jewish theoretical physicist whom Luria had encountered in Paris. Pontecorvo worked in Enrico Fermi's laboratory in Rome before Luria did, from 1932 to 1936, and published several papers with Fermi and his associates.[2] In 1936, Pontecorvo moved to Paris and worked at the Institut du Radium with Frederic Joliot-Curie. Luria met Pontecorvo when he came to the institute in 1938, and both men left Paris and arrived in the United States at about the same time. Pontecorvo spent two years in the United States, living in Oklahoma and working as a research physicist for the oil industry before he moved to Canada in 1943 to work at an Anglo-Canadian nuclear research laboratory at Chalk River, near Montreal.[3] In 1948, he took his family to Harwell, England, and continued his research at the Atomic Energy Research Establishment. A committed Communist since the Spanish Civil War, Pontecorvo defected to Russia after a brief holiday in Italy in late August or early September 1950.[4]

Pontecorvo had not lived in the United States for eight years before he ducked behind the Iron Curtain, and he and Luria had minimal contact while he was in Oklahoma. Nevertheless, Pontecorvo's defection alarmed the FBI, since "it [was] . . . possible to speculate that he may have betrayed reactor data from 1943 onward . . . thereby furnishing the Soviet Union with a particularly well-rounded picture of the Anglo-American-Canadian atomic energy work."[5] Quickly, the FBI drew up a list, which included Luria, of Pontecorvo's friends and acquaintances who might also be atomic spies. Thus, the FBI changed Luria's case from a "Security Matter—C" level to "Espionage—R, Atomic Energy Act," even though there was no reason to suspect him of espionage.[6]

On November 25, 1950, an internal FBI memorandum from agents in Washington recommended that Luria "be interviewed regarding his past association with Bruno Pontecorvo."[7] Although the current investigation "has failed to reveal . . . [espionage] activities," it "has determined that subject is in contact with known CP members."[8] The agents noted, "In view of the derogatory information set out above concerning subject, it is believed that investigation of subject should be continued after his interview, unless the interview clearly reflects otherwise."[9]

Two agents went to see Luria in his new office in the Noyes Laboratory at the University of Illinois. According to the report, "the interview was confined to BRUNO PONTECORVO and his known associates and contacts."[10] Luria provided the FBI agents with an account of his contacts with Pontecorvo in Italy in 1933 and France in 1938. He described how the two of them left Paris together by bicycle in June 1940, but became separated a day later. Luria and Pontecorvo saw each other briefly in Toulouse, France, but did not see each other in the United States until 1942. Luria told the agents that he did not think that he had been in contact with Pontecorvo after 1942.[11]

After questioning Luria about some of Pontecorvo's other European scientific colleagues, they asked about his recent disappearance. "LURIA did not attach any particular significance to the disappearance of BRUNO PONTECORVO. LURIA advised there did not appear to be great reason for concern over [his] . . . disappearance, inasmuch as . . . [he] had already given the Russians most of the data they needed."[12]

The administrative page of this report describes Luria's attitude toward the FBI agents. They report that "his general demeanor was one of complete friendliness" and "on the surface, he apparently endeavored to indicate a very sincere and frank attitude."[13] He told them that although other professors at the University of Illinois had asked him about Pontecorvo's whereabouts, he did not discuss the matter with them, because he "had no real knowledge of BRUNO PONTECORVO's activities and did not wish to take a chance on being sued for libel." Nevertheless, Luria "would endeavor to cooperate completely with any agent of the government as he felt it was his duty as a citizen to do so."[14]

Despite the fact that Luria seemed to have no information regarding Pontecorvo and there was no evidence that he was engaged in any sort of espionage or Communist activities, the investigation, including the mail cover, continued. Offices all over the country were directed to investigate the identities of Luria's correspondents, and the FBI seems to have monitored the telephone calls that were made to and from the Luria residence.[15]

Around the same time, agents from the Washington field office met with Ugo Fano, Luria's boyhood friend from Turin.[16] According to the FBI agents who visited him, Fano was "very evasive and refused to give a direct answer" when asked about Luria's loyalty to the United States. However, he allowed that he had heard that "the Curie Laboratory in Paris, France was Communist infiltrated; therefore, he is always somewhat suspicious regarding LURIA's association with the Laboratory in 1939 and 1940."[17] Fano also seemed to cast some doubt on Luria's sincerity in becoming a United States citizen, claiming that Luria gave the impression that his application for citizenship was "lip service" rather than a real desire to claim loyalty to the United States government.[18] Despite this "slight doubt in his mind regarding LURIA's loyalty to the United States," Fano had no reason to believe that Luria was in any way associated with the Communist Party or had been in Italy. Although another informant had claimed that Fano had told her that Luria had been ordered back to Italy by the Communist Party, Fano denied knowledge of any such orders. When the original informant was reinterviewed, she reiterated her claims that Fano had told her that Luria had appeared depressed after the war because he felt he should return to Italy, but the FBI agents did not pursue the issue with Fano.[19]

The investigation dragged on all through the winter and spring of 1951, although FBI agents acknowledged that their investigation had not been very successful. On a handwritten list of reasons to deny a request to interview more informants about Luria in January, the first reason given is "not warranted as investigation fails to indicate espionage."[20] The Springfield office report from May 1951 also had little of substance. A confidential informant who "is generally familiar with disloyal activities of individuals and organizations at the University of Illinois" had "been following the activities of the subject . . . for the past several months, and has not noted any indication that subject has been active in any activities inimical to the best interests of this country, including espionage activities."[21]

The confidential informant on campus apparently did not notice how, much as he had done when he first arrived on campus in Bloomington, Luria threw himself into university and local politics at the University of Illinois, Urbana-Champaign. He organized a local chapter of the American Federation of Teachers and became active in the local chapter of the Federation of American Scientists, a national organization of scientists committed to nuclear disarmament.[22] In addition, he joined a number of informal, ad hoc committees of professors and other civic-minded citizens to confront one of the most pressing issues facing the university: the debate over freedom of inquiry, loyalty, and censorship that came under the umbrella of academic freedom.

"Academic freedom" and "intellectual freedom" were the terms used in academic circles to refer to "issues raised by congressional investigations and [the dismissal or censoring of] Communist Party members" in academia.[23] This was of particular concern at public universities, where the institutions were financially dependent on local and state governments. In the spring of 1951, a group of Illinois professors, including Luria, met informally to discuss what they could do to protect their freedom of inquiry. At a meeting on April 12, they agreed, "It is our duty to a democratic society to affirm the principle of freedom of inquiry (academic freedom) at this time." Their responsibility to a free democratic society, rather than to the university or to a set of intellectual principles, prompted their statement. The professors proposed a statement of principle that outlined the democratic conditions necessary for intellectual freedom, and listed strategies for faculty in different parts of the large university to help them organize a coherent movement.[24]

The proposed statement of principle is a passionate analysis of the importance of freedom of inquiry and speech in a democracy. The authors note that in periods of crisis, "men tend to seek security by enforced conformity of thought and action." Intellectual freedom is suppressed "in the mistaken belief that thereby their security will be preserved." The authors argued that the opposite is true: that in times of crisis, clear thinking, cooperation, and consideration of all possibilities are necessary in order to maintain security. They boldly claimed, "Intellectual activity, when free and untrammeled, is a powerful dynamic factor and the ultimate source of strength and survival in a democracy. . . . Democracy and intellectual freedom are inseparable and indivisible." They acknowledged that with rights comes responsibility, and described the ways in which "teachers, scholars, scientists, artists, [and] writers" fulfill their responsibilities by educating students in such a way that "a democratic society moves steadily forward to higher levels of material and spiritual well-being for all of its members."[25]

The authors argued that "all forms of interference with intellectual freedom—loyalty oaths, censorship, restrictions on free speech, surveillance and intimidation" do not serve the "true interests" of democracy. These obstacles to academic freedom "devitalize and destroy the basic process from which a free society derives its greatest strength and its capacity to survive. No other form of society . . . has such capacity to release the potentialities of the human spirit and to marshal them in the service of mankind." It is unclear whom they held responsible for those obstacles, the government or the university administration, but the threat to American society was all-encompassing. Professors had a responsibility to protect "one of the precious sources of moral and material welfare which up to now have distinguished our form of society."[26] They called a university-wide meeting of the entire faculty to make sure that any statement they issued would reflect the views of as many Illinois professors as possible.

Their campaign moved slowly. In May, Luria was among a group of twenty-five professors who polled their colleagues to determine whether the faculty should proceed in developing a statement of principle.[27] Over the summer, 447 professors responded positively, and the informal committee met again in the fall to discuss how to proceed.[28] Although they had not intended the May letter as a referendum, they were heartened by the fact that only "20 members of the faculty indicated lack of interest in, or opposition to, the idea." The overwhelming sense from the positive responses was that it would be most appropriate for the American Association of University Professors to coordinate any faculty discussion and action, and so they turned the issue over to that group. The Faculty Senate also appointed a Committee on Academic Freedom, which took the lead in coordinating the faculty response to perceived threats. While the FBI knew about this movement among the Illinois faculty, and Luria's role in it, it did not merit additional surveillance or discussion about Luria's loyalty to the United States.[29]

Indeed, the FBI's claims that Luria was a Communist or subversive in some other way were rapidly losing strength. In December 1951, FBI agents once again visited Luria, this time to ask about another colleague. During that interview, Luria offered a statement about his own political beliefs. It is not clear whether Luria volunteered this information or if it was in response to a question, but he told the agents that he "considers himself a liberal who is willing to take the side of freedom against bigotry and to fight for equality of the minority."[30] This attitude did not raise any red flags for the agents making the report. (No pun intended.) In addition, a new informant reassured the FBI that "as a youth in Italy, [Luria] was a studious and well educated person who nurtured sentiments unfavorable to the Fascist Regime, [and] that his family is of good moral conduct and oriented towards the political parties of the right."[31] Although the formal investigation was closed on June 3, 1952, the FBI continued to monitor Luria for the next twenty years and to provide information about him to other government agencies.[32] The State Department was particularly interested in what the FBI had found.

"YOUR PROPOSED TRAVEL WOULD NOT BE IN THE BEST INTERESTS OF THE UNITED STATES"

In October 1951, Luria applied for a new passport for himself, Zella, and Daniel. Luria had been invited to give a lecture at the Society for General Microbiology meeting in Oxford, England, in April 1952, and afterward he and Zella planned to go on to Italy to visit with Luria's parents and introduce them to their grandson. Luria's passport had recently expired, and he filled out the forms and submitted two group pictures for a family passport. In November, his father Davide died, and Luria decided to visit his mother alone after the meeting in England. Luria duly informed the Passport Division of this change in plans and requested an individual passport.[33] Luria's attendance at the Oxford conference was announced in the scientific press, and he prepared his paper, an overview of his research on "The Mechanism of Virus Multiplication," for the printed proceedings of the meeting.[34]

In the early 1950s, Ruth Shipley, the chief of the Passport Division of the State Department, took it upon herself to protect the country from Communist and other subversive influences by regularly denying passports and visas to Americans and foreigners whom she deemed to be threatening. The McCarran Internal Security Act of 1950 dictated that known Communists should be denied passports, but Shipley went further and denied passports to anyone *she* suspected of being a Communist on the basis of her own investigations.[35] She justified her policy by saying, "One of the things I believe in is refusing passports to Communists. They've been working against us for a long time."[36] She was not required to explain her decisions, and the secretary of state publicly defended her as a fine civil servant.[37] Until late in 1952, there was no formal process by which rejected applicants could have their cases reviewed.

On January 25, 1952, Luria received a terse letter signed by Shipley. "After carefully considering your request, the Department is of the opinion that your proposed travel would not be in the best interests of the United States. In the circumstances, a passport is not being issued to you."[38] Luria immediately wrote back, saying, "I do not understand what the reason of the denial may be, but think it may involve some misunderstanding." He described in detail his invitation to speak before the Society of General Microbiology, "a distinguished recognition, as I am the only American bacteriologist invited to this year's meeting." He also discussed his plans to visit his mother, who "has been ill and dejected" since his father's death. He emphasized, "These are the only purposes of my trip. The denial of a passport would keep me from addressing the Society of General Microbiology and prevent me from visiting my Mother."[39]

Luria was aware of the potential political reasons for his passport denial, and he sought to ease any worries that the State Department may have had about his loyalty to the United States and the purity of his intentions.

> I wonder if the denial is connected with my past acquaintance with the physicist Bruno Pontecorvo, concerning whose activities I have been interviewed repeatedly by agents of the Federal Bureau of Investigation in the past several months. My acquaintance with Dr. Pontecorvo, however, is something of over 10 years ago. I saw him frequently in 1939–40 when we both worked in Paris, . . . just as I saw many other Italian scientists working there at that time. In this country I only remember seeing him once in New York in 1942, as I told the FBI agents in our conversation. At no time have I been in correspondence with him, and have long since lost contact with most common acquaintances. . . . I also wish to point out that at no time have I been engaged in, nor have had access to, confidential or restricted research. My own wartime work on government contracts was unclassified and dealt with bacterial resistance to antibiotics.[40]

Luria concluded the letter with a request for "an opportunity to submit any further information or evidence that may be desired, either by mail, or preferably, in person" in the hope that the passport chief would reconsider her decision to deny his passport.[41]

Shipley once again sent back a brief reply: "Your request for a passport has again been carefully considered but the Department does not feel warranted in providing you with passport facilities for travel abroad at this time."[42] Luria was stymied, and informed the Society for General Microbiology that he would not be attending the meeting in April.[43] In early March, he turned to University of Illinois president George Stoddard for help in a last-ditch effort to get a passport.

Luria showed Stoddard his correspondence with Shipley, and Stoddard was outraged. He sent a letter to Howland H. Sargeant, the deputy assistant secretary for public affairs at the State Department, demanding an explanation. Stoddard explained that Luria was "right at the top" of the community of microbiologists, and that he planned to go abroad to deliver a paper at a scientific meeting. Stoddard's "biological friends assured . . . [him] that it is a most distinguished company and that Dr. Luria deserves his place among them." Stoddard listed the dates of Luria's correspondence with Shipley, and then called Sargeant's attention to "certain interesting facets to this case." He reported that "Dr. Luria tells me flatly . . . that he is not a Communist and is not sympathetic to the Communist cause." As further proof of Luria's loyalty, Stoddard informed Sargeant that Luria, like all University of Illinois faculty, had "taken an oath of allegiance."[44] Although the copy that Luria signed was not included in the letter, Stoddard did attach a blank copy of the form (figure 3.1).[45]

Stoddard noted that the McCarran Act only allowed the State Department to refuse a passport to an individual who was known to be a member of the Communist Party or one of its fronts. Since "it is Dr. Luria's honest opinion that he belongs to no such organization," Stoddard and Luria wondered "if any organization he has ever belonged to has now been so classified." Stoddard reminded Sargeant, who was apparently his colleague in UNESCO, that the Universal Declaration of Human Rights includes the right to travel freely to other countries. Stoddard was frustrated by the apparent disregard for human rights by the State Department, and he was particularly upset by their lack of communication about Luria's case. "The Shipley memorandum is completely inadequate to Dr. Luria and to me; it makes a mockery of the ideals for which we both stand." He asked that the State

UNIVERSITY OF ILLINOIS
URBANA, ILLINOIS

The following affidavit must be executed and returned to Mr. C. C. DeLong, Bursar.

AFFIDAVIT

I, _____, do solemnly swear (or affirm) that I believe in and pledge my allegiance to the Constitution of the United States and the system of free representative government founded thereon; that I do not nor will I advocate the overthrow of the Government of the United States by force or violence; and that I am not a member of nor will I join any political party or organization that advocates the overthrow of the Government of the United States by force or violence.

_____
(Signature of employee)

Subscribed and sworn to before me this _____ day of _____, A. D., 19____,

at _____   _____
⎡ NOTARIAL ⎤   (City or place)   (State)
⎣   SEAL   ⎦

_____
Notary Public

M285

FIGURE 3.1
Loyalty Oath for University of Illinois employees, circa 1951. Photo courtesy of University of Illinois at Urbana-Champaign Archives.

Department either grant Luria a passport or "give evidence in support of its highly negative decision." Stoddard was prepared to take the case all the way to the president of the United States if the State Department were not willing or able to satisfy his demands.[46] He enclosed copies of Luria's correspondence with Shipley, and a document describing the "duty of the Department of State to Issue Passports."[47]

Sargeant wrote back on March 12. As soon as he received Stoddard's letter, he "discussed the case with the responsible officials of the Department." They and other officers of the State Department then reviewed the case, and they once again decided "under the spirit of the Internal Security Act" to withhold Luria's passport. Sargeant explicitly told Stoddard, "This decision is based on extensive and detailed security investigation reports, the contents of which cannot, unfortunately, be made available to you or to Dr. Luria." Under the circumstances, however, Sargeant was "satisfied . . . that a necessary decision" had been made. He was aware that his letter would not "lessen your indignation concerning what you consider an unjust and wise decision," but condescendingly told Stoddard, "until there is a change in national policy . . . we are forced to accept this decision."[48]

Stoddard shared Sargeant's letter with Luria, who was "indeed disappointed." In his reply to Sargeant, Stoddard reported that Luria was "not embittered" and was continuing his scientific research. Stoddard chastised Sargeant, and by extension the United States government, for broadly interpreting the McCarran Act as allowing this kind of secret condemnation to interfere with citizens' rights to travel. He emphasized Luria's feeling that he was "a loyal citizen . . . willing to meet any committee or agency empowered to check on his loyalty." Stoddard also communicated Luria's concern that the inability to travel would harm his scientific career, since he had already had to turn down invitations to other international scientific meetings, and "he now has no confidence in the willingness of the United States officials ever to let him leave this country."[49]

Stoddard was willing to pursue the case, since he was still upset about the State Department's violation of Luria's human rights. However, he respected Luria's request that Stoddard no longer try to help him, since Luria was not sure "he would be granted any such privileges of a hearing or an appeal." Stoddard limited himself to sending Sargeant a copy of an article from the *Daily Illini*, the student newspaper, highlighting the importance of Luria's work, perhaps with the hope that further evidence of Luria's stature in the scientific community would sway the State Department.[50] By late March, Luria was resigned to staying in the United States.

The FBI somehow discovered that Luria had made plans to leave the country. Hoover, seemingly unaware of the passport refusal, alerted the State Department Office of Security and Consular Affairs, the assistant attorney general, and the FBI legal attaché in London to the fact that Luria was going to England.[51] His work was unnecessary, and on March 26, the Springfield SAC sent Hoover an account of Luria's correspondence with the Passport Division, complete with quotes from his letters.[52]

On April 10, Luria wrote to his student James Watson telling him that he would not see him in England. "I guess you have imagined that the reason for the cancellation of my trip has been a passport difficulty of the type which is becoming increasingly common these days." He also asked the officers of the Society of General Microbiology "to refrain from any public statement."[53]

Luria's absence at the conference came at a crucial juncture in the history of molecular biology. At the meeting, Watson presented Luria's paper on virus multiplication, although Luria's ideas were soon eclipsed by the exciting results from Alfred Hershey that Watson also shared with the group.[54] Hershey and his colleague Martha Chase used different radioactive labels to tag the protein and DNA parts of bacteriophage. They demonstrated that even after "violent agitation" in a kitchen blender stripped off the protein coating of the virus, only phage DNA infected bacterial cells and "produced a normal crop of new phage particles," while the protein remained outside.[55] They therefore showed conclusively that DNA is the physical location of the genetic material, since it was the only part of the phage that entered the bacterium to take over its genetic mechanism to produce more phages. Watson, who was working with Francis Crick on deciphering the structure of DNA, was enthusiastic about Hershey's elegant experiment.

Caltech chemist Linus Pauling was also denied a passport in February 1952, and although he pursued a public appeal with the State Department, he was unable to attend a meeting at the Royal Society on protein structure that May.[56] Watson later mused about the effects of Pauling's absence on the outcome of the race to the double helix, hinting that if Pauling would have had an opportunity to see Rosalind Franklin's X-ray crystallography pictures of DNA while he was in London, he might have known how to interpret them as evidence of a double helical structure before Watson and Crick did so in February 1953.[57]

Luria's absence at the Society for General Microbiology meeting was noticed and discussed in hushed tones, but in accordance with his wishes, there was no mention of it in the official proceedings of the meeting.[58] In January 1959, Luria applied for a passport when he once again planned to travel to Europe for scientific and family reasons. He left blank the questions on the passport application about whether he was then or had ever been a member of the Communist Party but was nevertheless issued a passport.[59]

A PRODUCTIVE ACCIDENT

Despite Luria's personal and professional frustration over not being able to travel outside the United States in the early 1950s, his research program was unaffected. He continued to spend his summers at Cold Spring Harbor and traveled around the Midwest for regular phage meetings. He inadvertently became a fixture in laboratories across the country in 1951, when his postdoc Giuseppe Bertani published an article on his research on lysogenic bacteriophage. In the article, Bertani abbreviated a reference to his "lysogeny broth" medium for growing his viruses as "LB." The abbreviation soon became incorrectly known as "Luria broth," or "Luria-Bertani" solution, a staple in bacteriology and immunology labs to this day.[60]

The spring of 1952 was particularly productive scientifically. Luria and his graduate student Mary Human were still intrigued by the set of questions Luria had first explored a dozen years earlier with Max Delbrück: how does a bacteriophage particle replicate itself inside a bacterial cell? Luria was inspired by the work of Seymour Cohen, who used biochemical analysis to determine how "DNA synthesis immediately precedes and parallels the appearance of active phage particles . . . which suggests that DNA may be involved mainly in the final steps of the 'baking' of active particles."[61] Cohen's work led to another important question about bacteriophage replication: what happens to bacterial DNA while the cell is churning out viruses? Alfred Hershey also took a biochemical approach to understanding bacteriophage infection and found that bacteriophage DNA produces enzymes to break the bacterial DNA down into single units that are then reused when the cell produces more bacteriophage. Luria and Human chose to analyze the process of bacterial infection microscopically, and noticed that in most cases, the virus caused "more or less complete disintegration of the bacterial DNA."[62]

Those timed observations occasionally yielded some mutant *E. coli* that did not produce detectable bacteriophage T2 after infection. Luria had first noticed this phenomenon in 1946, when he found that the "juice" that remained after infected *E. coli* were killed sometimes did not reinfect other bacteria, but he did not pursue it at the time and "shelved it in . . . [his] mental files as 'the T2 mystery.'"[63] When he once again encountered this bacterial mutation, a laboratory mishap helped them understand this "novel situation."[64] The test tube of mutated bacteria Luria was working with broke, and as Luria later confessed, "I have never been a very neat laboratory worker, and this time the breakage proved to be a lucky break." He got a sample of different bacteria, *Shigella dysenteriae*, from Bertani and proceeded with his experiment, assuming that it would behave like the *E. coli* he had planned to use. "In fact they worked only too well. . . . My mutant bacteria had not failed to produce phage; they had produced a modified phage that refused to grow in its usual bacterial host, but grew perfectly well in Bertani's bacteria."[65] This strange result challenged one of virology's "most valid rules" about the relationship between viruses and their bacterial hosts: "that the properties of virus particles are unaffected by the host in which they grow."[66]

In a series of experiments with T2 bacteriophage and a range of *E. coli* and *Shigella* hosts, Luria and Human consistently noticed some of the bacteriophage disappear, only to reappear when cultured with different bacteria. One generation later, the same bacteriophage were once again able to reproduce in the original hosts. Luria and Human concluded that whatever caused the modification in the viruses was not a mutation, which would cause a permanent genetic change, or even evidence of the "peculiar plasticity of virus heredity," but rather evidence of "a new type of virus variation," a process they called "restriction-modification."[67] Bertani and others soon found similar examples of host-induced modification in other strains of bacteriophage.[68]

In April 1953, Luria's student James Watson and his Cambridge colleague Francis Crick announced that they had recently discovered the double helical structure of DNA, the molecule that constitutes genetic material. Fifteen years after Delbrück first thought to use bacteriophage as a model to probe the physical nature of the gene, bacteriophage researchers had provided a theoretical framework and some of the crucial experimental evidence necessary for describing DNA and its behavior. The close relationship between virus research and the structure of the double helix was further solidified in the months after Watson and Crick published their model in *Nature*. At the last minute, Delbrück added Watson to the list of speakers at the Cold Spring Harbor Symposium in Quantitative Biology in early June, which was on the broad topic of "Viruses."[69]

The excitement over Watson and Crick's elegant model dominated the symposium, although several other important conceptual and experimental topics were discussed over the course of the week.[70] One of those was Luria's paper on "Host-Induced Modifications of Viruses," the extension of the work he had begun the year before with Mary Human.[71] By the 1953 Cold Spring Harbor Symposium, Luria and his assistants, as well as researchers at other laboratories, had repeated the experiments with different strains of bacteriophage and several classes of bacterial hosts.[72] They consistently found that "the hereditary properties remain those of the virus, but they represent a core of potentialities on which the host can impose a new phenotype."[73] At that point, the best explanation for the "restriction-modification phenomenon" was that it was "due to the accident of acceptance of some particle of the restricted phage by some exceptional 'active' cell of the host." Luria and his colleagues "tentatively concluded" that the host-induced modifications somehow interfered with "some specific critical step of interaction with one or more hosts."[74] Luria urged the larger community of virologists to investigate whether the same effect could be duplicated in plant and animal viruses, since it could have significant effects on practical applications "in virus ecology and in the epidemiology and pathology of virus diseases."[75]

Those potential applications did not immediately materialize, since the phenomenon had only been observed in bacterial viruses. At the 1953 meeting in Cold Spring Harbor, there seemed to be no connection between Luria's observation and the new field of molecular biology. For the next decade, restriction-modification remained a question for virologists and bacterial geneticists, who were thinking about viruses as "bits of heredity in search of a chromosome," while molecular biologists focused on cracking the genetic code.[76] In the 1960s, the fields converged, when researchers discovered that the temporary changes Luria and Human had observed were the result of bacterial restriction enzymes that degraded viral DNA as a defense mechanism.[77] These restriction enzymes, which recognize and target short strands of DNA, are key molecular tools for recombinant DNA technologies that were developed in the late 1960s and early 1970s.[78] Recombinant DNA technology is the basis for genetic engineering and other types of genetic manipulation, with applications across biology and medicine, from agriculture to vaccine production. Luria acknowledged that his role in the history of recombinant DNA technology was "accidental" and "serendipitous," but as the technology developed he nevertheless felt an obligation to urge his fellow scientists to proceed cautiously with the molecular techniques he had a small part in discovering.[79]

BATTLING FOR ACADEMIC FREEDOM

In addition to his scientific work and his own struggles with the State Department, Luria continued to participate in on-campus activities to protect academic freedom. The biggest threat to academic freedom at the University of Illinois came from the state legislature, particularly from a state senator named Paul Broyles. In 1949, Broyles had proposed state laws that would require a loyalty oath from all state employees—university professors in particular—and set up a panel to investigate all subversives. The initial legislation was defeated by a veto by Governor Adlai Stevenson, but in the spring of 1953, Broyles reintroduced it as Senate Bills 101 and 102, and requested $65,000 for the new commission. The new governor, William Stratton, was "a conservative of almost extreme proportions," and a number of conservative Republicans had been elected to the state legislature in 1952.[80] The House Un-American Activities Committee had just begun its national "investigation of subversives in higher education," and so the climate was ripe for this type of anti-Communist legislation on a state level.[81] In Illinois, Broyles framed his proposals as "all-American bill[s] to defend our Constitution and our liberties."[82] In contrast to a loyalty oath that had recently been imposed in California, these bills did not specifically focus on university employees, but the statewide protest against the bills focused on their ramifications for academic freedom.[83]

The first of the Broyles Bills would establish a commission to investigate any individual suspected of Communist or other subversive activities against the government of the United States or of Illinois, with broad powers and little legal accountability. The other bill included a provision that required all state employees to swear loyalty to and prove citizenship of the United States and the state of Illinois.[84] Veterans groups strongly supported these bills and the opposition was dismissed as "Communist conspirators joined by fellow travelers, pinks, the misinformed and the uninformed."[85]

Those who opposed the Broyles Bills preferred to refer to themselves as the American Civil Liberties Union (ACLU), the League of Women Voters, the Illinois Chapter of the American Association of University Professors, the American Friends Service Committee, the Illinois Congress of Parents and Teachers, the Chicago and Illinois Bar Associations, the Chicago Urban League, and the Illinois Church Council.[86] Less formal groups, including a group of eighty-four mathematics professors and one hundred psychologists at the University of Illinois, also mobilized against the bills. Luria was an active member of the Champaign-Urbana Committee to Oppose the Broyles Bills, a group of local citizens, undergraduates, graduate students, and faculty members at the University of Illinois.[87] These groups sent letters to legislators, bought radio and newspaper advertisements, and sent letters to citizens throughout the state urging them to ask their state representatives to vote against the bills. The campaign stressed the unconstitutionality of the bills and emphasized the need for freedom of speech as a cornerstone of democracy.

The faculty senate of the University of Illinois, which Luria was a member of, passed a resolution condemning the bills in April 1953. The faculty made clear that they were "inalterably opposed" to Communism and Fascism, but were just as fundamentally opposed to the methods that Broyles proposed to protect the state from Communist influence. They questioned the need for an oath of loyalty to the state, and asked, "What considerations led to the conclusion that the State of Illinois, great though it be, could get along without the brains of persons who are not residents of this State? Illinois is limited, geographically. It must be unlimited intellectually."[88] The faculty declared the bills to be "a threat to the honored tradition of academic freedom." They warned that the atmosphere of suspicion created by the legislation would "jeopardize the ideals of freedom and justice cherished by all citizens of the United States."[89] The faculty senate sent its resolution to the university trustees and president, the governor, and members of the Illinois legislature.

The Broyles Bills, and the mistrust of academia that they reflected, were part of an ongoing clash between the University of Illinois and the state legislature over the presence of Communists on campus. In 1950, Representative Dillavou had claimed that there were "50 reds, pinks, and socialists" at the University of Illinois, a charge President Stoddard denied as vigorously as he had argued that Luria should be allowed to travel abroad. Stoddard acknowledged that the threat of Communists at the university was taken seriously, but tried to assure the legislators that the campus was free of enemies. "Those of us in charge have worked quietly, through our own security officers, the Federal Bureau of Investigation, the State Department and the military establishment, to make sure that no Communists are on the staff."[90] When pressed, the legislator had been unable to name a single such subversive on the faculty.[91]

In May 1953, Stoddard went to the state legislature with the university budget for the next year. During the debate over appropriations, Broyles asked Stoddard if he approved of the faculty senate resolution against the Broyles Bills, and Stoddard said that he did. Then Representative Horsley once again brought up the issue of Communists at the university. "I am sick and tired of the pinkos over at the University on the faculty . . . I am sick and tired of some of the profs and their pinko theories—and Dr. Stoddard isn't doing a thing to get rid of them."[92] Another representative wrote to Stoddard asking about the loyalty of the faculty in general, and the Committee of Mathematicians to Oppose the Broyles Bills in particular. She also asked Stoddard to swear "that the University taught, in all respects, that the American way is better than the Communist way."[93]

Stoddard continued to defend his faculty against the accusations of the legislature, but he did succumb to the pressure to "root out Communists." As opposition to the Broyles Bills mounted across the state, university administrators quietly informed Norman Cazden, an assistant professor of music, that his contract would not be renewed because they had evidence that he was a member of a Communist organization. Cazden left Illinois at the end of the spring term and was later subpoenaed by the House Un-American Activities Committee.[94]

Stoddard's efforts to protect his faculty while appeasing the legislature by expelling a Communist were not satisfactory to the university's board of trustees. On the night of June 24, they passed a vote of no confidence in him and demanded his immediate resignation over a series of problems, including the allegation of Communists on campus. Over the protests of Illinois students, faculty and staff, and numerous citizens, Stoddard left Urbana.[95] Luria was away at Cold Spring Harbor for the summer, but he instructed his friend Irwin Gunsalus to "sign my name . . . to any statement that you may consider proper to sign concerning the actions of the Board of Trustees at their last meeting."[96] In September, Luria wrote directly to Stoddard, calling his dismissal "an alarming sign of the times. They are part of the present climate of intolerance and bigotry, of anti-intellectualism and of aggressive mediocrity."[97]

The dismissals of Cazden and Stoddard did not appease the state legislature, which passed amended forms of the Broyles Bills in June.[98] Several revisions were necessary, since Governor Stratton was concerned about the amount of money allotted to the anti-Communist commission. He vetoed the bills in July because the loyalty requirement was so broad that it would place a burden on state resources "which . . . is completely disproportionate to the danger that might be involved."[99] Communists could be lurking everywhere, not just in the education system, and although he was sympathetic to the need for loyalty from all state employees, he felt that the legislation as it stood would not accomplish its goals.

Heartened by this close campaign, Broyles introduced his legislation for the third time at the next meeting of the state assembly in 1955. At the urging of the Chicago office of the ACLU, the Champaign-Urbana Committee to Oppose the Broyles Bills swung back into action.[100] Luria's graduate student Seymour Lederberg was the chair of the Student Committee on Discrimination and Academic Freedom, and Luria was a liaison for the local chapters of the American Federation of Teachers and the Federation of American Scientists.[101] Once again, they collected signatures on statements against the bills, circulating letters and fliers explaining why they were dangerous. Luria was one of six faculty members who published an article in the local newspaper urging their colleagues to join the ACLU and support the efforts to protect individuals from the "hysteria which is leading to repression."[102] The committee sponsored a public meeting in April, which featured local politicians, academics, and clergy members.[103] The faculty senate again made a strong statement against the bills, and criticized the new university president, Lloyd Morey, for his announcement of a new policy to punish those members of the faculty who invoked the Fifth Amendment when questioned about their political beliefs.[104]

This time, Broyles was able to protect his legislation from its opponents and from budgetary constraints.[105] On July 20, 1955, the administration announced that all university employees were compelled to sign a new loyalty oath and file it with the university bursar. In contrast to the broad language in the old oath, which had been in place since 1941, this one specified that the signer was "not a member of nor affiliated with the communist party and . . . not knowingly a member of nor knowingly affiliated with any organization which advances the overthrow or destruction of the Constitutional form of government of the United States or of the State of Illinois, by force, violence or other unlawful means."[106] Because the legislation was effective immediately, university employees had no choice but to sign their oaths right away in order to receive their July paychecks. Although the actual documents are no longer extant, we can assume that even the faculty who opposed the bills, like Luria, signed the affidavit and remained employed.[107]

CONFRONTING INSTITUTIONAL RACISM

The summer of 1955 was difficult for other political reasons as well. Earlier in the year, Luria was elected to the chairmanship of Section N (Medicine) of the American Association for the Advancement of Science (AAAS), which automatically gave him a position on the AAAS Council.[108] His responsibilities included organizing the program for this large section at the December 1955 meeting, scheduled to take place in Atlanta, Georgia. The council had chosen Atlanta as a host city in 1953, as an acknowledgement of "the veritable revolution in nearly every aspect of science and society" in the South.[109] One aspect of society in the South had not yet undergone a revolution, though, and the continuing practice of racial segregation posed a significant practical and ideological problem for the organizers of the large scientific meeting. Hotels, restaurants, and taxis in Atlanta conformed to Georgia's laws separating Blacks and whites, and the local organizers had found few facilities where AAAS members of all races would be welcome. Confronting institutional segregation revealed an ongoing, subtle tension in the scientific community over how to balance a scientific commitment to meritocracy and equal access to knowledge with discriminatory social and political policies. Luria and his fellow AAAS leaders wrestled with the question of how to appropriately protest a system that so blatantly violated their personal principles as well as the spirit of scientific inquiry and education.

In April, Section H (Anthropology), led by the distinguished African American physical anthropologist W. Montague Cobb, canceled its program at the Atlanta meeting to protest the segregated facilities.[110] Other council members expressed concern about the segregated conditions in Atlanta, and the AAAS board of directors met in June to assess the situation. The board issued a statement detailing the "Advantages of an Atlanta Meeting" in the June 24 issue of *Science*, the official AAAS publication.

> Because the Association recognizes no distinction of color in the achievement of its purposes, the situation in Atlanta was carefully explored. It was determined that all program sessions, the exhibits, the Science Theater, and the Association's social functions . . . could all be held on a non-segregated basis. Although all the professional activities are thus non-segregated, it unfortunately remains true that segregation still obtains in hotels, restaurants, and transportation. The board believes, however that scientists of all races will benefit from participation in this meeting, and that the advantages outweigh the disadvantages. . . . Several individuals and associations have urged the whole Association not to meet in Atlanta. The decision to meet there was made in 1953 after a most careful study of all of the factors involved, and the board believes that the original reasons are valid.[111]

The only other reference to segregation before the meeting was a statement in the July 22 issue of *Science*. The "Hotel Headquarters and Housing" list noted that, "Because of existing local laws, separate hotel and motel accommodations are named for Negro members and visitors."[112]

Despite the board's reassurances that the segregated conditions in Atlanta would not interfere with the mechanics of the meeting, some council and association members remained concerned about whether it was appropriate for a scientific organization, committed to freedom of inquiry and communication, to implicitly condone this segregated system. Race relations and the dismantling of segregation were increasingly sensitive national issues in the aftermath of the *Brown v. Board of Education* ruling of 1954, and fourteen-year-old Emmett Till's brutal murder in Mississippi in the summer of 1955 compelled many Americans to confront the devastating effects of white supremacy. Luria, his fellow council members, and the scientists they represented had a vigorous correspondence throughout the summer and fall of 1955 dealing with the same difficult questions facing the entire country.

In August, Luria sent a "personal and confidential" letter to Cobb about his hesitations about going to Atlanta. While he reiterated that the letter was not "an official communication from Section N," Luria spoke both as the chairman of Section N and as a scientist. "I am personally shocked as a member of [the] NAACP, and am also hesitant to assure my speakers that the meeting will be as unsegregated as the local committee has apparently promised." Despite the recent statement made by the AAAS board of directors, Luria "began to wonder if this step [of canceling the meeting], however distasteful, should not be considered seriously."[113] Luria also communicated with Dael Wolfle, the AAAS administrative secretary, and Warren Weaver, the chairman of the board, expressing the same discomfort.[114]

In their correspondence, Luria and Cobb discussed the possibility of polling all of their fellow council members to see if there was sufficient support for canceling the meeting.[115] In September, anthropologist Gabriel Lasker, secretary of Section H, wrote to Cobb and Luria, as well as to Harold Plough, secretary of Section F (Zoological Sciences), and Barry Commoner of Section G (Botanical Sciences), proposing two resolutions for the council to adopt before the meeting in December.[116] He remained concerned that if they attended the meeting without protesting the segregated conditions, "we shall fail to make it clear that we believe that in science the race of the investigator is of no consequence." The proposed resolutions asked council members to vote on whether or not the AAAS should cancel the meeting and also on a statement "That the AAAS never in [the] future arrange meetings at any place where members would be legally barred from full participation by reason of race."[117]

Luria was hesitant to sign on to the first resolution. After "serious thought," he told Lasker that he felt that "a referendum of the Council on cancellation of the Atlanta meeting at this time would be inappropriate, besides having little chance of producing a positive vote." Wolfle and Weaver had been "interested" in the possibility of a poll, but he agreed with them "that the arguments in favor of proceeding . . . however debatable, are strong." Nevertheless, he was "wholly in sympathy" with the second resolution, and proposed that it be circulated among the council members before the meeting so that they could discuss it in Atlanta.[118]

The issue continued to weigh on Luria's mind all through the fall of 1955. In October, he wrote to George Beadle, council president, and to Weaver, once again expressing his personal hesitations about going to Atlanta, especially "after what happened recently at the [Emmett Till] trial in Mississippi," when Till's murderers were acquitted by an all-white jury. He still felt that "as a matter of discipline within the Association" he would "go through with the program," but he "hope[d] that the Board of Directors and the council will see to it that the AAAS does not by their silence appear to condone discrimination in any form." Luria suggested that a "strong statement from the Board of Directors . . . would make all participants feel less uncomfortable in going to the meeting."[119]

Just as Luria wrote to Beadle and Weaver out of personal conviction mixed with professional responsibility, their responses reflect their personal feelings on the matter as well as their thoughts on the responsibility of the AAAS. They both reiterated the board's commitment to the Atlanta site and in different ways expressed exasperation with the continued discussion over whether to hold the meeting. Beadle's response was brusque. "Sure, I know what you mean," he wrote. "None of us likes segregation. And the Board made this clear in *Science*. No AAAS function is segregated but we can't change the state laws and city ordinances." He preferred to lead by example, and wondered, "What would we accomplish by not holding a meeting in Atlanta??—very little it seems to me."[120] Weaver was more dismissive:

> I did not suppose, when I went to a AAAS meeting in Alaska, that I was by that procedure expressing approval of the sub-zero weather that occurs in that part of the country; nor have I ever considered that my visits to India imply an approval of the Hindu attitude towards cows and monkeys, nor of Nehru's politics. In other words, it seems to me farfetched to suppose that the fact that the AAAS chose to meet in Atlanta implies any approval of some of the disgraceful episodes which have occurred, in the southeastern part of the United States, with respect to segregation.[121]

Despite Weaver's and Beadle's attitudes, other members of the AAAS continued to express their discomfort with Atlanta, and some were critical of the way in which the board was responding to their concerns. Quentin Young, a Chicago physician, sent Luria a copy of a letter he had sent to Wolfle along with his annual dues. Young was irate that the AAAS would choose to go to Atlanta, "the city where the full force of law enshrines the foremost anti-scientific social lie of our age, the myth of white racial superiority." In his response to Young, Luria assured him that "Personally, I feel very much in agreement with your position" and that as an officer of the AAAS, he was committed to "represent[ing] as faithfully as possible the opinion of the members in the Council." The Section N members who had written directly to Luria were all "strongly against holding meetings in segregated cities."[122]

In early December, the Illinois section of the Society for Experimental Biology and Medicine voted to contact George Beadle to express their disapproval of the choice of Atlanta, and their worry that "It may be interpreted by some that the Association tacitly approves segregation." They encouraged Beadle to institute a policy to ensure that no future meetings would be segregated.[123] The Chicago section of the American Association of Clinical Chemists made a similar plea both to their parent organization and the AAAS.[124] The discussions continued up until the meeting began on December 26. Several members of the council proposed statements to be discussed and voted on at the council meetings on December 27 and 30.[125]

The issue was framed as one of scientific integrity, rather than a moral or legal matter. An organization dedicated to scientific ideals could not be seen to be contradicting those ideals by allowing its activities to be curtailed in any way by segregation. For example, Bernard Davis's proposed resolution took the AAAS's stated goal of "the effective use of science in the promotion of human welfare" as a starting point for his argument. He felt that the council members would agree that "the attainment of our goal is hindered by attitudes or measures that interfere with the full opportunity of all individuals . . . to develop and utilize their talents."[126] The council committee on resolutions synthesized the proposed statements into one that was vigorously debated at the meeting. Ultimately, 224 council members voted by mail to adopt a statement that reaffirmed the AAAS's commitment to full participation in the scientific process by all of its members:

> In order that the Association may attain its objectives, it is necessary and desirable that all members may freely meet for scientific discussions, the exchange of ideas, and the diffusion of established knowledge. This they must be able to do in formal meetings and in informal social gatherings. These objectives cannot be fulfilled if free association of the members is hindered by unnatural barriers.
>
> Therefore may it be resolved that the annual meeting of the American Association for the Advancement of Science be held under conditions that make possible the satisfaction of these ideals and requirements.[127]

A few weeks after he returned from Atlanta, Luria tried to put a positive spin on the episode. He wrote to Cobb, "I feel that through the unfortunate decision made several years ago of meeting in Atlanta real progress has been made in the thinking of the great majority of the members of the council and that such unfortunate decisions will unlikely be repeated in the future."[128] The AAAS did not meet again in the Old South until 1990.[129]

LEAVING ILLINOIS

Luria was actively looking for a new position in the mid-1950s. Conflicts with the state legislature aside, he was content in the bacteriology department at the University of Illinois, but strict university rules against nepotism prevented Zella from securing a permanent faculty appointment. This widespread policy, established during the Depression to ensure that as many families as possible had an income, meant that only one member of a family—nearly always a man—could be on the employed by a university.[130] Although the rules protected universities from political pressure in making appointments, it also stifled the careers of many female academics.[131] Zella earned her Ph.D. in 1951 from Indiana University, but over the next several years she was only awarded occasional one-semester teaching jobs, for very little pay, in the psychology department at the University of Illinois.[132] If Zella were to have an academic career of her own, the Lurias would either have to fight the university administration or move.

Fighting for a job was not a viable option. In the early 1950s, the University of Illinois administration renewed its commitment to enforcing the nepotism rules, after relaxing those standards during World War II. The rules were quite strict. There was "absolute restriction" on first-degree relatives both appointed to the academic staff, to the point where a female tenured math professor was not reappointed after she married an untenured member of her department in 1953.[133] Thanks to efforts by the American Association of University Women, and to a lesser extent, the American Association of University Professors, these rules were eventually repealed, but not in time to benefit the Lurias.[134]

They first spent the 1958–1959 academic year on sabbatical in Boston, where Luria had a visiting professorship at the Massachusetts Institute of Technology (MIT). That year, he was approached by the University of Wisconsin–Madison to replace his friend Joshua Lederberg, and the Albert Einstein School of Medicine in New York, where he could be the chair of the microbiology department.[135] Almost as soon as he arrived in Boston, the MIT administration proposed making him a permanent member of their biology department. Luria did not want to move to New York but wrote a terse note to Max Delbrück: "Complex choice for me: MIT, permanent, or Madison, or Urbana. I'll probably take the first manned satellite."[136] When Zella received an offer from the Tufts University psychology department, they decided to remain in Boston permanently.

When Luria notified the University of Illinois of his plans, university officials pleaded with him not to leave. They wrote to Zella, outlining the provisions they might be able to make for her in the psychology department, but their priority was to keep her husband, not to support her career. To Luria, they emphasized that they needed him there "for your scientific eminence, for your educational counsel, and for your influence on graduate students."[137] In February, Luria wrote to Dean Lyle Lanier that their decision to move to Boston was final. He tried to reassure him that "the reasons for my choosing MIT are purely personal; the opportunity for me to develop a biology training program of an unusual kind, and the opportunity for Zella to pursue a career of her own in a completely separate institution."[138] After considering the various towns around Boston, the Lurias bought a home on Peacock Farm Road in nearby Lexington, where it was easy for both of them to commute to their jobs in Medford and Cambridge. There was a large community of academics and a high-quality school system for ten-year-old Daniel.

4 RECOGNITION AND RESPONSIBILITY

TRANSFORMING BIOLOGY AT MIT

Luria's move to MIT afforded him an opportunity to develop new aspects of his scientific leadership, as he was tasked with guiding the biology department through a significant change in focus and personnel during a time of dramatic change in the life sciences. His talent as a teacher and influence as an advisor far surpassed the significance of the research he did in this period, due in large part to his talent for attracting promising students and researchers to MIT, and his sense of where the next big research projects and funding would come from. Basic biological research became increasingly oriented toward cancer in response to political interest in the disease in the 1960s, and Luria was well aware of the fact of American scientific life that basic research had to be justified in terms of practical applications in order to be politically appealing.[1] The American Cancer Society had funded Luria's research since he arrived in the United States, and so he was in a position to take a leadership role in articulating the scientific and administrative tasks needed to explore the link between biological processes and cancer.

In the late 1950s, the biology department at MIT was at a crossroads. As many full professors reached retirement age, department and university administrators felt that it was a good time to reevaluate the goals of the department of biology and the broader agenda of life sciences research at the institute.[2] In the fall of 1956, acting department head Irwin Sizer and institute president James Killian asked Warren Weaver, the vice president of the Rockefeller Institute, to chair a special panel charged with assessing the strengths and weaknesses of the department of biology.[3] University administrators and faculty members alike noted that while the current department's focus on general physiology, biochemistry, and biophysics reflected the basic science and engineering ethos of the institute as a whole, there were emerging areas of the life sciences that MIT should emphasize in order to build a first-rate department. In the immediate post–World War II climate, biophysics at MIT had been promoted as "a wholesome alternative or anti-dote to nuclear physics" and institute officials emphasized it as the core of the biology department.[4] But when it came time to restructure the depart-ment in 1956, administrators took note of the state of biological science and sought researchers in other "fields which could profit best from the MIT environment," and microbiology replaced biophysics as the future of life sciences research.[5]

In addition to the department evaluation by biologists, Victor Weisskopf and others on the physics faculty sensed that the nascent field of molecular biology was going to grow into a fruitful and lucrative research area, and they "prodded" the administration "to make the Department of Biology a center for research and teaching in molecular biology."[6] They urged the administra-tion to recruit high-profile geneticists and molecular biologists in order to "modernize" the undergraduate and graduate programs. When Luria agreed to join the permanent faculty in 1959, it was a major step in achieving those goals.

The need to reform MIT's department of biology reflected a larger trend in academic science in the United States from the late 1950s through the 1960s. In the pages of *Science* and the *American Institute of Biological Sciences Bulletin*, biologists and educators debated about the future of the life sciences in the wake of the molecular revolution. Biology departments had tradition-ally been split between botany and zoology, with genetics and embryology in separate departments, and with little contact among these distinct groups of researchers. Bacteriology and microbiology were often part of a university's medical faculty, and undergraduates were not regularly exposed to these dis-ciplines. All areas of biology were faced with the opportunities and perils of "going molecular" and the accompanying shifts in boundaries between traditional areas of inquiry.[7]

Observers feared that undergraduates had historically seen biology as an easy science, and worried about the ways in which many institutions and departments narrowly focused on preparing students for medical school.[8] Moreover, shifts in the teaching of high school biology meant that by the mid-1960s, many college freshmen would already have been exposed to the ideas that were presented in introductory biology courses.[9] These internal disciplinary challenges were matched by the post-Sputnik political pressure on the scientific community at large to maintain American superiority in all branches of science.[10]

The future success of biology as a broad discipline seemed to be in jeop-ardy. Despite the recent rapid growth in molecular biology, one biologist observed, "We teach people the wisdom they need to solve yesterday's problems instead of tomorrow's." He lamented the fact that as a mature science, "biology . . . has everything but sophisticated instruction about how to become a biologist."[11] Several biologists and educators found "the lag period between classroom biology and the research frontier . . . appalling."[12] The old disciplinary distinctions were no longer useful for students or researchers. The solution seemed to be an academic restructuring. Throughout the 1960s, "on campus after campus, traditional departments of botany, zoology, and bacteriology were merged to form departments of biology entailing a complete overhauling of introductory courses and major realignments of power."[13] The process that brought Luria to MIT was an early part of this national trend away from traditional approaches to biology.

When restructuring the department of biology, Sizer also had to consider the teaching and research needs of the institute. At the time, the department averaged forty undergraduate biology majors per year, but Sizer sought to increase that number. He felt that the department should expand in a limited number of areas "in order to develop an outstanding reputation by specialization is a few fields, rather than by distributing a small teaching faculty among different subjects." Although the current undergraduate curriculum did not need major changes, he recommended that many of the courses offered be "modernized" in the next few years. He also emphasized the importance of providing funding for talented graduate students.[14]

Above all, though, Sizer felt that the best way to improve the department would be "to build an outstanding faculty" in biophysics, biochemistry, and analytical biology, and to invite visiting professors in other "special areas." He was eager "to get rid of the dead wood of the department" and fill it with the finest researchers in molecular biology.[15] As an individual who was "one of the two or three people who, through their research and leadership, have made virology what it is today: a vigorous branch of biology, contributing in important ways to the branches of genetics, biochemistry, radiobiology and immunology," Salvador Luria was on the top of his list of potential faculty members.[16] Luria was invited to be a visiting professor in the 1958–1959 academic year.[17]

On December 12, 1958, Weisskopf sent a note to president Julius Stratton urging him to recruit Luria for the good of the department and the institute. "He is not only an unusual and outstanding biologist . . . but he is also one of those few very outgoing and extravert personalities who spread their influence way beyond their field of specialization. This is just the type which we need so badly at MIT."[18] Weisskopf went to introduce himself to Luria and wrote another note to Stratton on December 16, reminding him that MIT had the financial resources to match Luria's promised microbiology laboratory at Wisconsin. The personal contact only strengthened Weisskopf's conviction that Luria was the right person for MIT, and he enthusiastically told Stratton, "I have received an even stronger impression that this is exactly the man we need. He is a fairly dynamic person who really wants things to be done on a large scale; at the same time he is a thorough and genuine scientist with deep understanding for the philosophic essentials. A rare combination, but here it is. . . . All I want to say is that here is an opportunity which we almost cannot afford to lose. If we get him I am sure that MIT will be at the forefront of biology."[19] Luria's appointment would be a great boon to MIT, since it meant that the biology department could focus on microbiology, a topic that "comprises all that is interesting and loaded with future developments" and "fits in so admirably with the new objectives."[20]

It was an opportunity Luria did not want to lose either. The chance to revitalize the department of biology and set up a full microbiology laboratory was appealing, and he was encouraged by the institute's commitment to making microbiology the center of undergraduate training.[21] Luria's friend Cyrus Levinthal had been hired the previous year, and he was pleased with the supportive environment.[22] Moreover, after a few short months, the Luria family found Boston to be comfortable politically, socially, and intellectually.

The members of the department of biology were "all delighted" when Luria decided to stay. Their delight was practical as well as intellectual, since a researcher of Luria's stature had the ability to bring significant financial resources to the institute.[23] Sizer anticipated that outfitting a new microbiology laboratory for Luria would cost the institute approximately $46,000 to pay for air conditioning, structural changes, furniture and teaching microscopes, and additional personnel.[24] He proposed that Luria apply for a National Science Foundation grant for the $100,000 needed to build him a proper lab. Although there was "no assurance" that Luria would receive the grant, Sizer was confident that it was "very likely."[25]

When Luria permanently joined the MIT faculty, he asked for a multilingual secretary. On September 15, 1959, Nancy Ahlquist came in for an interview. Luria asked her three questions. He wanted to know if her children would interfere with her career, if she would have trouble with his accent, and if she could start that Monday.[26] This was the beginning of a twenty-five-year working relationship and close friendship based on mutual admiration and respect. Luria had nothing but praise for her, remarking that she made his "professional life possible and easy. She has also been a model of a superior person, helpful, straightforward, and willing to put me in my place when I became discourteous or unreasonable."[27] On her part, many years later she commented that others had told her that she "would either love Salva or hate him, but I never found out what I was supposed to hate."[28]

The process of transforming biology at MIT was not complete once Sizer was named department head and Luria was hired. On February 15, 1959, Sizer sent Stratton a handwritten memo about the tasks ahead. Now that Luria had agreed to join the faculty, Sizer wrote, "We are now ready to go on to the next problems of space, financing, a General Microbial Physiologist (Magasanik?), a Mammalian Cell Biologist, etc."[29]

Boris Magasanik was actively recruited from Harvard's department of bacteriology and immunology once Luria had agreed to stay at MIT, and Luria's presence there was the most important factor in his decision to move to the institute. He recalled that he was "strongly attracted both by the prospect of working in close vicinity to an outstanding microbial geneticist such as Luria and by the plan to modernize together with him and other faculty members the undergraduate and graduate programs."[30] The hiring continued, and in the fall of 1959, president George Stratton reported that in addition to Luria and Magasanik, biochemists Vernon Ingram and Philipps Robbins were joining the department.[31]

VIRUSES AND CANCER

In 1959, Harry Weaver invited Luria was to give the "keynote and orienting address to outline the present situation with regard to the viral etiology of certain tumors" at a three-day meeting organized by the American Cancer Society (ACS) for "experienced investigators" to discuss "The Possible Role of Viruses in Cancer." The organizers suggested that he present "a series of the most penetrating questions that he can frame, so that the other speakers (and future researchers) can concentrate on them."[32] Luria and his colleagues were asked to be "brutally objective" in their evaluation of "problems that seemingly must or should be solved, [and] . . . the nature of those obstacles that appear to be impeding future progress."[33]

Luria had a long-standing relationship with the ACS. The private organization had generously funded Luria's research from the time he arrived in the United States even though his work was not directly related to cancer therapeutics. For the better part of the twentieth century, the ACS program was behind a broad trend in basic biology funding to tie "pure biology" research to the problem of cancer.[34] From the time that Peyton Rous first identified the chicken sarcoma virus in 1911, virologists had sometimes doubled as cancer researchers, searching for the cause of human cancers. In addition to this targeted search for the direct cause of cancer, the ACS supported basic research on genetics, viruses, and other biological processes through its Committee on Growth, since there was tantalizing evidence that viruses were somehow implicated in tumor development.[35] By studying normal cellular and genetic function, the ACS and its scientists reasoned, they would better understand how cancer cells behaved differently.

Funding basic research gave the ACS a constant source of news items to demonstrate to the general population that their donations really were helping scientists cure cancer.[36] For example, in January 1952, Patrick McGrady, the science editor of the ACS wrote a press release about Luria's research. The report began with the dramatic announcement that "University of Illinois scientists have uncovered one of the vital steps in virus reproduction within the cell and may have opened the way for new methods of controlling viruses, including those which cause animal cancers." After describing Luria's latest research results with virus reproduction and latency, McGrady reminded readers "while science has uncovered about a half-dozen animal cancers which are virus-caused, there is no evidence as yet of viruses causing any human cancers."[37] Nevertheless, these reports were powerful fundraising documents and often included a request for additional donations.

Luria was always careful about the claims he made with regard to the relationship between his research and cancer. In an application for additional funds to the Illinois Division of the ACS, he described his proposed experiments on virus latency in detail, acknowledging that "the bearing of these problems on cancer is an indirect and hypothetical one," but nevertheless "at the center of the whole question as to whether virus-like agents may be involved in the etiology of tumors from which infectious viruses cannot be recovered."[38] The ACS apparently agreed with his reasoning, and they continued to fund his research.

The 1959 conference brought together a diverse group animal, plant, and bacterial virus researchers, including Luria, his old virology antagonists Peyton Rous and Wendell Stanley, his former collaborator Renato Dulbecco, and French molecular biologists Francois Jacob and André Lwoff. In his address, "Viruses, Cancer Cells, and the Genetic Concept of Virus Infection," Luria took as his starting point "recent advances in basic virology, which have produced some unifying concepts in the relation of viruses to cellular constituents and to cellular functions."[39] Cellular functions, in turn, were fundamental to understandings of cancer as the result of abnormal cell growth and control. He argued that viruses could be implicated in the development of cancer either as the cause of somatic mutations or as a direct infection of a tumor-causing agent. In both cases, "virus infection may be considered as a class of cellular mutations: a class of mutations . . . in which . . . the primary change, the entry of the viral genome is a genetic change." From that assumption, the idea of studying animal tumors as an instance of "infective heredity at the cellular level" was a viable one to pursue.[40]

Luria was not arguing that cancer was caused by viral infection but rather that research on viruses as external causes of cellular change would yield important results for understanding cancer. In the conclusion to his remarks, he made that point quite clear:

> Personally, I am inclined to believe that the majority of cancers stem from genetic changes affecting long established cellular elements rather than from the entry or activation of relative newcomers such as full-fledged viruses. Yet, I also believe that the most rapid progress in cancer research, considered as a branch of cellular genetics, will come from the study of virus-induced tumors, because they provide us with the opportunity to analyze the role of individual genetic elements in the transformation of normal cells into tumor cells.[41]

In addition, evidence from bacteriophage infection of bacterial cells, as well as from other virus-host examples, showed that gene transfer played a significant role in cellular alterations that could characterize not only viral infection but tumor growth as well.[42]

As Weaver had hoped, Luria's review sparked a vigorous debate.[43] Luria and other bacteriophage researchers understood the link between viruses and cancer as based on a conception of "virus infection as a class of mutations," which in turn "may be responsible for one or several of the recognized steps of carcinogenesis."[44] Others, such as C. H. Andrewes of the National Institute for Medical Research in London, felt that this definition was too broad, and could not see how virological or pathological research on virus infection would provide any insight into tumor growth.[45] Nevertheless, enough virologists, geneticists, and biochemists were convinced that tumor viruses held the key to understanding and controlling cancer that over the next ten years, both the American Cancer Society and the National Cancer Institute gave significant amounts of money to support research along these lines.[46]

ADMINISTRATOR, TEACHER, MENTOR

Luria played a key role in attracting research grants, graduate students, and other researchers to the department as he and Magasanik immediately began to plan the reorganization of the graduate biology program around microbiology. "Assuming that the task of biology as a science is to explain the organization of living matter at all levels . . . we propose to develop the structural and functional approach emphasizing simplicity and unity of pattern rather than complexity and diversity." Using microorganisms, they hoped to introduce students to the properties of all living matter before giving examples of the various solutions to "the problem of maintenance of biological organization under specific evolutionary circumstances." The emphasis on microorganisms would parallel the undergraduate program and provide research and laboratory support for introductory courses.[47] The presence of eminent microbiologists and a strong research program attracted many eager students, and the graduate community expanded rapidly in the early 1960s.[48]

As anticipated, Luria quickly brought increased funding and positive attention to the institute. In January 1960, the National Institutes of Health awarded MIT a $100,000 grant to construct "laboratories for research into the borderlands of life."[49] The institute's press release focused on Luria's work on the role of viruses in cancer and emphasized the importance of a strong program in cellular biology.[50] Luria was elected to the American Academy of Arts and Sciences in 1959 and to the National Academy of Sciences in April 1960, and his reputation continued to attract other molecular biologists and bacteriophage researchers to MIT.[51] In the fall of 1960, Luria was delighted when David Baltimore arrived in Cambridge to begin his doctorate (although he completed it at the Rockefeller University), and the virologist James Darnell joined the faculty to run the tissue culture-virology program.[52] Luria had invited Darnell to help him revise *General Virology* for its second edition, and they had fourteen years of new research to incorporate into the text.[53] When the work was finally complete, it was warmly reviewed and praised again for its comprehensive scope and accessibility for undergraduates.[54] By 1964, more than a dozen new faculty members, including Maurice Fox and Eugene Bell, had arrived at MIT, and the department had more than doubled the number of research associates.[55]

Luria took on his fair share of teaching undergraduates and graduate students. In the fall of 1961, the department introduced a new curriculum in the life sciences intended to train students not only for careers in medicine but also all aspects of biological research. This program was so popular that by the fall of 1962, "the teaching facilities of the Biology Department ha[d] been placed under considerable strain."[56] Ahlquist recalled that Luria was a popular teacher whose reputation spread quickly across campus: his enrollments went from 75 students to 150 and then 300 over the span of three years.[57] This was paralleled by an overall increase in courses and students enrolled in the department.

The new undergraduate program in biology conferred a degree in quantitative biology at the end of four years of study. The MIT course catalog reminded students that this was a "new approach to the biological sciences" and that the biology that they would learn had "evolved . . . from the conventional fields of classical botany and zoology." The new biology "is an exact science, utilizing physics and chemistry and surpassing both these fields in its scope and its immediate importance to man's life and health."[58] Although this catalog description remained constant throughout the 1960s, the number and diversity of the courses increased significantly. Luria and Magasanik taught courses in microbiology, Cyrus Levinthal taught biophysics and molecular genetics, and by 1970, seminars in virology, biochemistry, molecular biology, and protein and nucleic acid structure were offered to advanced students.[59]

In 1968, Luria began to teach Biology 7.01, the introductory undergraduate general biology course, in addition to his advanced courses in microbiology. This was the first time he had taught such a course, although he had been "for many years a client of the introductory biology course."[60] As Luria taught it, basic biology "centered around one main theme, that of living organisms as possessors of a program—a set of genetic information—that underlies all vital functions and that evolves by mutation, genetic recombination, and natural selection."[61] Genetics, developmental biology, and physiology were the disciplinary foci of the course, while many of his examples came from bacterial and phage genetics. He justified the exclusion of such fundamental areas as zoology, botany, evolutionary theory, and ecology by saying that he felt that those areas "are best approached after a general biology course in which the student learns the essentials of life phenomena: the chemistry of the cell, the organization and function of the genetic systems, the genetic significance of life cycles, and the functioning of cells in differentiated organisms."[62] In addition, Luria felt that because of the rapid advances in biology at the time, it was imperative for "practitioners of experimental biology . . . [to] address . . . the beginning biology student with a voice from the laboratory."[63] Genetics and molecular biology had made "the structure of biology . . . uniform and explicit" and so any overview of the field had to focus on those areas rather than on evolutionary theory or zoology. The evolutionary significance of genetic information was a large part of the course, but in trying to emphasize "molecular . . . and evolutionary unity," he was committed to "the persistent use of genetic concepts."[64]

Luria's approach to introductory biology was so successful that he continued to teach the course even after he was named an institute professor in 1970. In 1975, the last year that he taught 7.01, the MIT Press published *36 Lectures in Biology*, which consisted of Luria's lectures and classroom illustrations from the 1973–1974 academic year. The book records how Luria divided the course up into five sections: cell biology and cell chemistry,

biochemistry, genetics, developmental biology, and physiology. Throughout the course, he emphasized the interplay between mechanical and phe-
nomenological interpretations of biology in order to make the point that
evolutionary developments are not teleological.[65] He cautioned his students that as they continued to investigate biological phenomena, their
"desire to . . . see what hides inside should be balanced with the reverence
that everything rare and precious deserves."[66]

Luria admitted that he enjoyed teaching undergraduates more than
graduate students, whom he expected to learn in the laboratory rather than
in the lecture hall or seminar room.[67] He nevertheless took a keen interest
in their lives and intellectual development and was particularly concerned
about "the growing prejudice of some . . . [graduate] students against the
world of non-science."[68] To counter that prejudice and to foster his own
interest in literature, Luria encouraged his new graduate students to read
"broadly and purposefully," and he and Zella hosted a weekly Sunday night
literary seminar to discuss the "possible relevance of biological concepts
to humanistic studies."[69] He chose a theme that was relevant for both science and the humanities, such as "Man in Nature and in Books" or "The
Existentialist Approach to Conflict," and identified books from history, philosophy, poetry, and fiction that would help him and his students grapple
with these topics.[70]

These stimulating Sunday evening conversations complemented the
discussions about "The Life Sciences and the Humanities" that Luria was
engaged in as a member of the Salk Institute's Council for Biology in Human
Affairs.[71] Luria had been selected as one of the original nonresident fellows
of the Salk Institute in 1966, and he appreciated the institute's commitment
to fostering relationships between scholars who shared a common interest
in exploring the role of science in the modern world.[72] He brought insights
from those discussions back to MIT as an active participant in the institute's
Technology and Culture Seminar, an ongoing forum for MIT faculty to "discuss the meaning and impact of their work" in a larger social and cultural
context.[73]

BOSTON AREA FACULTY GROUP ON PUBLIC ISSUES

Luria felt very comfortable in Cambridge, with its liberal political atmosphere. Early in the 1960s, along with fellow MIT faculty members Levinthal
and Noam Chomsky and Harvard professors Hilary Putnam, George Wald,
Everett Mendelsohn, and several others, he coordinated efforts by scientists

and other academics to shape the public debate over nuclear weapons testing and other related issues.[74] Operating under the umbrella name Boston
Area Faculty Group on Public Issues (with the unwieldy acronym BAFGOPI),
they formed smaller ad hoc committees to respond or call attention to political issues such as civil rights, disarmament, nuclear testing, civil defense,
and education. By early 1962, BAFGOPI had ninety-three members.[75] This
cadre of academic activists distanced themselves from the group of Cambridge intellectuals who shuttled back and forth between Harvard and the
White House, advising the president on economic and military matters. If
Arthur Schlesinger Jr., Irving Kristol, and John Kenneth Galbraith represented
the elite intellectuals, Luria, Mendelsohn, Putnam, and Chomsky saw themselves as the public's intellectuals, who chose to communicate their ideas
about national policy directly to the American people.[76]

In 1962, the steering committee of BAFGOPI defined the aims of their
informal group and invited colleagues from around Massachusetts to join
their efforts. They emphasized that their voices were crucial parts of any
democratic system, especially during the critical episodes of the Cold War. By
virtue of their specialized knowledge in the natural and social sciences, and
the traditional openness of academic communities, the steering committee
felt that they were in a special position to teach both the government and
the country. Their desire to educate the public was an outgrowth of "the natural function of teachers to disseminate truths as they see them, and to dispel ignorance and misinformation." Although in the past, "convention has
restricted this process primarily to the classroom and academic journal . . .
the complexity of our times and the fateful approach of possible catastrophe
present to us a wider responsibility." The "truths as they saw them" led the
BAFGOPI leadership to adopt principles that would allow them to use their
social and cultural authority to speak out on "all major issues directly relating to the preservation of peace with freedom and the prevention of war."[77]

That spring, BAFGOPI recruited eight scientists for a formidable task:
responding to physicist Edward Teller's public advice about how to prepare
to survive nuclear war.[78] In February 1962, the *Saturday Evening Post* ran
three excerpts from Teller's book *The Legacy of Hiroshima*. Teller, a participant in the Manhattan Project and a visible political presence, was a vocal
advocate of preparedness as a way to win a nuclear conflict with the Soviet
Union. Teller's insistence on building up a nuclear arsenal rather than nego-
tiating for disarmament generated considerable concern among the Boston professors. They welcomed the opportunity to join a small but vocal
group of scientists who were questioning "the extent to which democratic

procedures were being subverted by the failure of scientists to provide the public with facts and information sufficient to allow their full participation in political debates about the wisdom of atomic testing" and nuclear preparedness.[79]

The *Saturday Evening Post*'s editors, "believing that every citizen should be given the opportunity of judging for himself the merits of conflicting opinions in this vital controversy," gave Teller's critics the opportunity to express their views. Luria, who had been one of the original signatories to Linus Pauling's 1957 open letter from scientists calling for a nuclear test ban, volunteered to participate.[80] He and biologist Matthew Meselson joined six physicists and engineers from Cornell, Harvard, and MIT who responded in "An Answer to Teller."[81] Their passionate and angry article was an attempt to refute Teller's proposals on scientific and practical grounds. They argued that Teller's "Plan for Survival" was "a signal example of the combination of factual error and emotionalism which might lead to . . . catastrophe." The authors accused Teller of abandoning "the cautions customary to scientific reasoning" in his proposals for how to survive and win a nuclear war.[82] They criticize Teller's position as "unrealistic and unsound" for dismissing the utility of negotiated arms control, underestimating the catastrophic effects of nuclear war, advocating a costly and ineffective shelter program, and placing an "unjustified trust" in a nuclear defense system.[83]

In their estimation, Teller was not acting according to scientific reasoning, since he based his analysis on "what may be possible rather than on what is probable," a strategy that the eight authors deemed "madness."[84] With evidence of widespread damage from nuclear tests, Hiroshima, and biological research on radiation and mutations, they countered Teller's rosy picture of the American ability to survive a nuclear attack, and argued that the only scientifically justified course of action was a realistic arms control agreement. The eight authors concluded by urging the American citizen, with whom the ultimate power lies, to "keep a firm grip on sense and reality and turn his back on fear and madness."[85]

By positioning themselves as spokesmen for science, Luria and his fellow BAFGOPI authors were able to offer an alternative to Teller's scientific authority and give a more hopeful and positive image of what could be accomplished through scientific reasoning. Teller's plan was "not only an illusion, but also a tragic dissipation of all hope for the future."[86] In contrast, their proposal for arms control would ensure the survival of the world. Their perspective became part of the national debate over the future of the nuclear arms race as their message was read by thousands of American citizens.

THE RESPONSIBILITY OF BIOLOGISTS

Luria was quite busy with public discussions of science that spring. In April, he was invited to present at a special symposium on the "Control of Human Heredity and Evolution" organized by his former Indiana colleague Tracy Sonneborn.[87] Biologists from a range of disciplines met on April 6 on the occasion of the dedication of a new biology building at Ohio Wesleyan University. The papers and some of the discussions were later published as a series of five essays in a slim volume intended "to arouse public awareness . . . of the increasing need for intelligent and wise public judgments and actions" on the emerging possibility of control of human heredity and evolution.[88] Luria's essay, "Directed Genetic Change: Perspectives from Molecular Genetics" contained a discussion of "The Responsibility of Biologists," which emphasized what scientists must do to ensure that their work is used for the good of a free and democratic society.[89]

Luria began his analysis with general comments on science and society. Because "no science is completely pure since its findings always bear, however indirectly, on human affairs," all scientists have "an inescapable responsibility" both to find useful applications and to avoid any "line of research that is clearly leading to evil applications."[90] The scientific community as a whole must acknowledge that "science creates power" and that they must therefore always be sure to communicate their findings and any consequences or applications to the general public in order to avoid creating a technocracy. Scientists should share the responsibility for deciding how to use the results of their research with well-educated citizens. This is especially important for researchers in rapidly growing fields such as biological engineering, since "the progress of science is often so rapid . . . that it creates an imbalance between the power it places in the hands of man and the social conditions in which this power is exerted." Scientists thus have "an additional, more subtle responsibility" to be alert to the possibility of this rapid expansion, and maintain open communication so that the public will be prepared for the subsequent advances in knowledge.[91] According to Luria, in order to fulfill this responsibility, researchers may have to abandon their "habits of skepticism and restraint, of curbing fantasy and distrusting fancy," which had previously prevented them from confronting the possible useful and harmful uses of radioactivity.[92]

Luria's paper for the conference focused mainly on trends in biological research that could someday lead to the control of human genetics through molecular and other biological engineering, through what he called "genetic surgery." He took the time to raise this issue of scientific responsibility not because he thought that direct control over human heredity was rapidly approaching, but because he felt that this branch of science would have such a major impact on the "basic issues of human values and public policy" that it was crucial to consider these issues early.[93] If the infrastructure of science included provisions for this type of communication and education, then when fundamental issues arose there would be a social and political context in which to evaluate them. Any laboratory advances would have to be accompanied by social and cultural change on the part of the scientific community and the general public.

Luria returned to the issue of responsibility in the conclusion of his presentation. He was not confident that the changes in genetics would confer any special new responsibilities on geneticists, but he also did not want scientists to ignore the possibility of rapid expansion of that science. The specter of Nazi eugenics, and other lessons of the first half of the twentieth century, made him cautious when thinking about the possible applications of "genetic surgery." He reminded his fellow scientists that we can "no longer . . . believe that the growth of knowledge and control over our physical environment will itself guarantee that the resulting power will be utilized wisely." Scientists should not "subside into fatalistic despair or withdraw into the ivory tower of social agnosticism." Rather, they should "create some machinery by which the social implications of our work can be debated rationally and openly."[94]

At this point, Luria believed that the responsibility and role of science in general and biology in particular were to improve human existence and to protect basic human rights. Scientists could not pretend that their work was neutral, and they had a special responsibility to make sure that, through rationality and reason, science would become (or remain) a source of good in the world. There was no need for government control of science, but there was a need for intensive communication between scientists and citizens. Public education was key, whether the issue was nuclear weapons or genetic engineering. Luria was confident that this approach would not restrict or limit science in any way. As he commented at the end of the conference, "Once genetics makes apparent both the tremendous dangers and the great potentialities for good, people may become more alert to, and more willing to accept, the enormous promises of this science."[95]

CHANGING SCIENTIFIC DIRECTION

With the high demand for his teaching at MIT and his public engagement on scientific issues, Luria was ready to take a one-year leave in 1963, once "the microbiology outfit . . . [was] in good running shape," in order to explore a new field of research.[96] For several years he saw that the research that most bacteriophage workers were doing in molecular biology "consisted in putting together little pieces of a large puzzle whose overall features were already evident," and felt that he was "too old to be anything but a boss and a specialist."[97] The pace of molecular biology research was too fast for his tastes, and he was increasingly intrigued by the role of enzymes in bacterial cell membranes.

Luria was introduced to the topic when Japanese bacteriologist Hisao Uetake spent some time in his laboratory in Illinois. Uetake and his colleagues had found a strain of bacteriophage that altered the surface of the *Salmonella* bacteria they were infecting. Different strains of *Salmonella* bacteria can cause typhoid fever and foodborne illnesses, and the strain-specific polysaccharides (sugar molecules) on their cell membranes are the antigens that prompt the human immune system to produce antibodies to the bacteria. While he was in Luria's lab, Uetake found that some bacteriophage permanently changed those antigens. He and Luria determined that somehow the bacteriophage DNA was "bringing into a bacterium the genes for new cellular components," which had exciting implications for understanding how viruses might be involved in "the transformation of a normal cell to a cancer cell."[98] The "search for the possible phage enzymes that converted the bacterial antigens" would require complex biochemical analysis that they were not trained to do.[99] Uetake had to return to Japan and Luria moved to MIT before they could find a biochemist to help them unravel this phenomenon, but Luria invited Uetake's colleague Takahiro Uchida to join his laboratory in 1960 to continue their research.

Luria's colleague Phillips Robbins was searching for a way to understand how cells synthesize complex sugars (such as antigens), and Uchida helped him learn how to work with the unusual bacteriophage and *Salmonella*. Robbins eventually described not only how antigens are converted but also the way that enzymes in the bacterial membrane build those antigens as well. Luria was intrigued by the enzymes Robbins had identified and was particularly interested in learning more about the cytoplasmic membrane of bacteria, which is the only organelle in a bacterial cell.[100] In order to learn as much as he could about cell membranes, he proposed that they start what they called the Microdermatology Project to host seminars on bacterial cell surfaces. That international reading and discussion group introduced Luria to colicins, a class of proteins produced by bacteria that kill other bacteria by attacking the surface of their cells. Although colicins have the same effect on bacteria as bacteriophage, studying their biochemical effects required

Luria to develop an entirely new set of laboratory techniques. Luria was awarded a second John Simon Guggenheim Memorial Foundation Fellowship, this time to spend a year in Paris, at the Institut Pasteur, so he could turn his undivided attention to this new research program.[101]

The year abroad passed pleasantly for the Luria family. First, Luria spent two weeks teaching at the Weizmann Institute in Israel before they settled in the Latin Quarter of Paris. Zella took a leave from her position at Tufts, and she and Dan quickly became fluent in French. Just as he had done in 1938, Luria spent the year learning biochemical techniques and obtaining enough results to generate a series of productive research questions to explore once he returned to MIT. He established professional ties and personal friendships with his French molecular biology colleagues, particularly Jacques Monod, and found the "courage" to follow an artistic impulse and take sculpting lessons with Piera Rossi.[102]

AWARDS AND RECOGNITION

When he returned to Boston in the fall of 1964, Luria was named MIT's first William Thompson Sedgwick Professor of Biology. This professorship honored Sedgwick, a bacteriologist who had been the first head of the department of biology in 1889.[103] Awarding the chair to a prominent microbiologist and geneticist made a clear statement about the scientific priorities of the department of biology and the institute administration. With his new title and new research program, Luria settled into the next phase of his career.

In June 1969, Luria and Delbrück were selected as the recipients of that year's $25,000 Louisa Gross Horwitz Prize awarded by Columbia University Medical School, for their work on bacterial genetics.[104] Established in 1967 to recognize excellence in basic research in biology or chemistry, the Horwitz Prize quickly became regarded as a predictor for Nobel Prizes. More than half of the winners have also been awarded a Nobel Prize—some, like Luria and Delbrück, within weeks of collecting their awards.[105] On October 8, the two met in New York to deliver a short lecture at a formal dinner in their honor.[106] A week later, they received the news that they were sharing the Nobel Prize in Physiology or Medicine with Hershey "for their discoveries concerning the replication mechanism and the genetic structure of viruses."[107]

The Nobel Committee's explanation of why they chose Delbrück, Luria, and Hershey described their broad influence on the life sciences and medicine, and noted that their work established bacteriophage as a model for animal and human cells. The committee credited them with providing the foundations for the "explosive development" of molecular biology and "a deeper insight into the nature of viruses and virus diseases," along with relevant findings for genetics and the study of development and growth. The summary concluded, "Over the years our debt of gratitude to the three leading figures of bacteriophage research has continually increased."[108]

This Nobel Prize was given in recognition of a lifetime of work, not for "the kinds of spectacular breakthrough discoveries for which the prize is almost always given."[109] As David Baltimore wrote, Delbrück, Luria, and Hershey were honored as "the prophets, conscience and experimental innovators of molecular biology."[110] Other bacteriophage researchers, most notably Luria's friend Joshua Lederberg and his student James Watson, had been already been recognized by the Nobel Committee in 1958 and 1962 for spectacular breakthroughs—the discovery of bacterial genetics and the elucidation of the molecular structure of DNA, respectively—research which rested on the foundation Luria, Delbrück, and Hershey had created. Lederberg immediately wrote to Luria, "1958 made no sense at all before 1969. Maybe I can honestly begin to enjoy it now; the student the better when the teacher has been honored too."[111]

FIGURE 4.1
Salvador Luria receiving the Nobel Prize, December 10, 1969. Photo courtesy MIT Museum.

Most of the press coverage and scientific response emphasized the importance of bacteriophage research for molecular biology, although Luria's MIT "co-workers say that almost all of the advances in control of viral diseases in recent decades have stemmed at least indirectly from the research of Drs. Luria, Delbrück, and Hershey."[112] In addition to outlining in *Science* the quality and influence of their research, Gunther Stent praised the trio for their integrity and leadership in making "molecular biology not merely a nice place to visit but also a good place to work."[113]

The State Department agreed that Luria's travel was in the best interests of the United States, and he flew to Stockholm with Zella and Dan to receive the Nobel Prize (figure 4.1). Luria was pleased and proud when both Prime Minister Olaf Palme and King Gustav VI preferred to talk to Zella than to him during the festivities.[114] At the Nobel Prize ceremony on December 10, Professor Sven Gard, a representative of the Karolinska Institute praised the trio's bravery in taking on an "overambitious" goal of explaining "the most fundamental of all biological problems" by using "the lowly bacteriophage." Although this approach "probably . . . raised many eyebrows," he credited their "sense for the importance of strict scientific methodology, . . . brilliant experimental skill and above all . . . imaginative approach" with their "success in making the impossible feasible."[115]

In his Nobel lecture, Luria described how he was currently using bacteriophage and colicins "as probes into macroregulatory phenomena of the bacterial cell," investigating how genes regulate "some of the major processes of the living cell," particularly the permeability of cell membranes.[116] He and his students were in the process of understanding the biochemical process by which colicins interfere with protein and DNA synthesis, as well as the function of the cell membrane. Luria compared the state of colicin research to the early stages of bacteriophage research, noting that in both cases "the use of simple bacterial systems represents a departure from the traditional materials of the respective disciplines, genetics and 'membranology.'" Enthusiastic and optimistic about the new community of researchers he was creating, Luria was hopeful that he would once again be a part of a research program that produced "something meaningful and exciting."[117]

5 PROTESTING THE VIETNAM WAR

"AD-ITORIALS"

When the Lurias returned to Boston from Paris in the fall of 1964, the American mood was grim. The country had been rocked by the assassination of President Kennedy in November 1963, the violent response to the passage of the Civil Rights Act, and episodes of race-based police brutality in the summer of 1964.[1] The country stood on the precipice of entrenching itself in a full-scale war in Vietnam after the passage of the Gulf of Tonkin Resolutions in the summer of 1964. Although the U.S. military had been involved in Vietnam for several years, President Johnson significantly increased manpower and firepower in Southeast Asia in the winter of 1964–1965.[2] Early in the war, the rhetoric defending the escalation resembled the typical Cold War posturing used in the 1950s to justify U.S. involvement in Korea: the United States had to protect South Vietnam from Communist rule, lest the Chinese and the Russians overrun the entire country from the north.

Opposition to the war began slowly, with different left-leaning political groups adding their voices of dissent.[3] When the United States military involvement in Vietnam escalated with bombing runs after the Gulf of Tonkin Resolutions of 1964, Luria along with many others felt that the United States had entered a morally indefensible situation. Although BAFGOPI had considered dissolving itself early in 1964, the organization was strong and able to mobilize quickly.[4] The Boston peace community was made up of several overlapping groups, and Luria was also a member of the Massachusetts Political Action for Peace (MASSPAX) advisory committee.[5]

Many existing groups that had mobilized to speak out against nuclear weapons now turned their attention to the U.S. military involvement in Vietnam. The protests took many forms, but most often they involved mass demonstrations in Washington, D.C., or teach-ins on college campuses.[6] Luria participated in many of these events in the Boston area, but he chose to focus his efforts on a more specific form of protest in order to reach a national audience. BAFGOPI organized a number of ad hoc committees to sponsor advertisements in national newspapers, especially the *New York Times*, the *Boston Globe*, and the *Washington Post*, to convince their fellow citizens, the government, and the world that the U.S. involvement in Vietnam was wrong.

In effect, these advertisements served as paid editorials, signed by hundreds. Throughout the Vietnam War, a flood of these ads appeared in the pages of national and local newspapers, and although some historians have claimed that they "had no discernible effect on any Administration's policy," they nevertheless became a tool for public protest that is still used by academics and other public figures today.[7] Government agencies, newspaper publishers, and American citizens were introduced to a new mode of communication in the free market of ideas, open to anyone who had the $8,000 to pay for it. By 1967 this type of protest was so recognizable that advertisements of this sort were dubbed "ad-itorials" by *Newsweek* magazine.[8] BAFGOPI members took credit for starting this trend, jokingly boasting that if faculty and students at the University of Michigan took credit for inventing the teach-in, "We invented *The New York Times*."[9]

While Luria deeply appreciated the democratic freedoms that allowed him to use the press to criticize the U.S. government, there was also a practical reason why he focused on this form of protest. At this point in his life, Luria was suffering from sciatica and other back problems. He was eager to join the thousands of others at protests in Washington and "let myself be arrested," but his MIT colleague and fellow war protester Noam Chomsky "warned me that even one night on a steel plank in a D.C. jail such as he had experienced would cause my poor twisted spine, an ailment I have suffered from since I was forty years old, to immobilize me for a month."[10]

The largest and most famous ads that BAFGOPI's Ad Hoc Committee on Vietnam sponsored appeared on January 15 and 22, 1967, at a point when public frustration with the war was high. On two successive Sundays, a full page of the *New York Times* Week in Review section was devoted to a simple message. "Mr. President: STOP THE BOMBING." No moral arguments, no explanations, just an imperative and a list of over 2,000 names of academics from more than 80 colleges and universities, as well as a number of clergy, authors, artists, and educators from the Children's Museum of Boston.[11] The ad announced that "additional signatures from more than 140 other Colleges and Universities in over 27 States have been received and will appear in the *New York Times* Sunday January 22." The second ad listed thousands more names (figure 5.1).

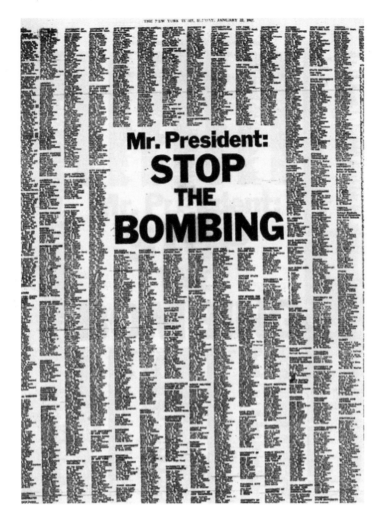

FIGURE 5.1
"Mr. President: STOP THE BOMBING" ad, *New York Times*, January 22, 1967.

On January 21, Luria, Harry Lustig, a physicist at City College, along with Morton Fried of Columbia's anthropology department and Harvard philosopher Hilary Putnam, held a press conference at the Overseas Press Club in New York about the ad. They announced that at that point over 6,000 faculty members at 200 colleges in 37 states had signed the petition and contributed to the cost of publication. Luria and Putnam explained that they had asked the signers to pass the petition along to their friends, and thus had collected the signatures in a "haphazard" way. The collection of names was not "intended to be a poll—we think there are many, many more people in the academic community who, for one reason or another, would support the plea."[12]

Despite the simplicity of the message of the advertisement, Luria and his friends made several points to the press. They claimed that although they were mostly Democrats or independents, they were frustrated by the Democratic administration's lack of interest in pursuing negotiations. Luria and Putnam noted that there were several new Republican lawmakers whom they felt would be more receptive to messages asking for alternatives to bombing. The professors told reporters that they hoped that the United States was not persisting in bombing North Vietnam as a way to avoid the possibility of negotiations. "In a family quarrel . . . neither wants to give in, but somebody has to take the first step to end the dispute. That we continue the bombing is one position that prevents negotiations," Luria explained.[13] Luria and Putnam were trying to take the part of the wise relatives who compromised their position for the sake of a peaceful resolution.

While Luria enjoyed "the machinery of soliciting, gathering, and publicizing" the signatures of fellow academics for these political advertisements, he also took a number of opportunities to argue against the war as a private citizen.[14] For example, in a rare moment of embracing his Jewish identity in 1968, Luria published an essay in the *Boston Jewish Advocate* entitled "Babi Yar, Warsaw, and Vietnam," which compared American silence over Vietnam to German complicity in the Holocaust.[15] As "a Jew who left Europe to escape Nazi persecution and found in the United States a haven and a new fatherland to love and respect," he identified with the Vietnamese civilians who were attacked with napalm and whose villages were destroyed by bombs. Luria placed the responsibility for these atrocities, which he called pogroms, with American military and political leaders, not the American soldiers who were carrying out flawed orders. He reminded his readers that under Hitler the German people did not have the same rights to assembly, a free press, and fair elections that American citizens enjoy, and so he urged them to use those rights to oppose an unjust war. "History will deal . . . severely with us Americans if we shall not have used our free voices, our free press, our free votes to put an end to the crimes perpetuated in our name!"[16] Many readers rejected his comparison between Vietnam and the Holocaust, and questioned whether it was appropriate for Luria to speak on a topic that had nothing to do with science.[17] Nevertheless, Luria felt that he was justified in taking this position as a Jewish refugee from Nazi Europe, and was gratified when the article was reprinted after he won the Nobel Prize in 1969.

RESISTING A CALL TO SCIENTIFIC ARMS

In the summer of 1967, biophysicist Ernest Pollard wrote a letter to the editor of *Science* magazine, expressing surprise that the scientific community had not rallied around the United States government the way it had during World War II, and calling on his fellow scientists to support the Vietnam War. He had "no desire to influence anyone's opinion but . . . would feel better if a segment of American science at least were actively seeking to improve the U.S. position in Vietnam." He suggested that "scientific aid" would help the U.S. military counteract terrorism and "permit humble people . . . to choose their way of life without fear for themselves or their families." He asked scientists to write to him to sign up to be a part of this cadre of scientific advisors, and he promised to report the results in the pages of *Science* and send the names along to President Johnson.[18]

Luria and his friend and fellow immigrant scientist Albert Szent-Györgyi responded swiftly to Pollard's proposal. They sent off a letter to the editor of *Science* calling Pollard's view "a false and distorted one." Luria and Szent-Györgyi noted that many of their scientific peers had been among the thousands of professors who had signed public statements and newspaper ads opposing the war, and encouraged their peers not to respond to Pollard's invitation. They bluntly called the war "a national catastrophe and a moral blight for our country" and encouraged other scientists to "consider, carefully . . . whether cooperation with the Johnson Administration in waging the Vietnam war is consistent with service to the true interests of our country and of mankind." According to Luria and Szent-Györgyi, scientists had an extra responsibility for the well-being of mankind, regardless of political ideology. They concluded their letter by urging those scientists who had come out against the war to join them in "review[ing] our present professional activities in order to make sure that they do not unnecessarily contribute to the waging and prolongation of that war."[19]

Their letter was published in *Science* on October 6, and two weeks later Pollard described the response to his initial call. Pollard was pleased to report that 179 individuals, from high school students to a college president, wrote to him volunteering time to support the war. Individuals also made suggestions as to how the scientists could contribute, including studies on pacification, health issues, and biology. Many respondents "concluded that dissent had gone too far," and Pollard himself felt that "the schism among scientists caused by the present situation is latent with profound tragedy." Pollard was encouraged by the response to his letter, but he acknowledged that they had problems of "placement, of briefing and of budget."[20] The editors of *Science* balanced his report with several critical letters that echoed Luria and Szent-Györgyi's points. Some suggested that the best way for the American scientific community to respond to the war was to encourage its end, and pointed out that there was already considerable scientific investment in the technologies and weapons used in the war.[21]

Luria and Szent-Györgyi received a few private responses to their letter, including an admiring letter from a college student in Vancouver with a request for advice on the social responsibility of the scientist.[22] The editors of *Science* chose to publish a number of harsh responses that attacked their expertise, authority, and patriotism.[23] Biologist Stanley Buckser took Luria and Szent-Györgyi to task for "step[ping] outside their fields of competence to advise other scientists not to help their country in time of national emergency." He argued that many of the antiwar critics had voted for Johnson, and that "these people should search their own consciences before advising others not to help their government in augmenting the policies that they voted for them to make." E. Staten Wynne of the U.S. Air Force School of Aerospace Medicine challenged Luria and Szent-Györgyi's claim to speak for the scientific community, arguing, "Dissenters are more likely to express their views than supporters." He encouraged scientists who supported the war "to make sure their professional activities do just that" in order to avoid having to fight Communism closer to home.[24]

In order to prove the point that most scientists in fact supported the war, Wynne used an example that involved Luria. He noted that in the spring of 1967, the American Society for Microbiology (ASM) overwhelmingly voted down a proposal "to dissolve the standing committee advisory to the U.S. Army Biological Laboratories." At that same meeting, Luria was installed as the president of the ASM. The relationship between this scientific society and the U.S. Army laboratory at Fort Detrick, Maryland, was one of the defining issues of his term in office.

THE COMMITTEE ADVISORY TO THE UNITED STATES
ARMY LABORATORY AT FORT DETRICK

Fort Detrick was designated a U.S. Army Biological Laboratory during World War II. Wartime projects included research on anticrop agents (which led to the development of Agent Orange) and infectious biological weapons. During the Cold War, this program continued and Fort Detrick expanded into a large, modern facility, with laboratories researching botulism, smallpox, and anthrax as potential biological weapons. By the time of the escalation

of the Vietnam War, army biologists and epidemiologists had tested different methods for spreading bacterial and viral agents and stockpiled strains
of several deadly diseases.[25] In the midst of their research, those scientists
turned to outside sources in the scientific community for general advice
and guidance in technical matters.

In 1955, the Chemical Corps of the U.S. Army asked the Society for American Bacteriologists "to appoint a committee to act in liaison between the
Society and the Chemical Corps" and advise them on microbiological questions.[26] (The SAB renamed itself the American Society for Microbiology in
1961.) The committee provided advice on scientific and professional matters;
it had "no responsibilities regarding moral, political or military questions."[27]
Members of the committee were appointed to two-year terms by the president of the ASM, and several presidents served on the committee at various
points in its history. The committee members had security clearance from
the army "for access to SECRET defense information," and they met once a
year at Fort Detrick for a two-day meeting with the scientists and physicians
on staff. The committee's scientific advice provided the technical director of
Fort Detrick "with an evaluation of the planning, execution, analysis, and
interpretation of . . . scientific projects," based on discussions with researchers "regarding specific current research topics such as microbial genetics, tis-
sue culture, pathogenesis of infectious disease, immunology, etc."[28]

Some members of the ASM were uncomfortable with this relationship,
which seemed to condone the development of biological weapons. At the
joint council and Council Policy Committee (CPC) meeting on April 29,
1967, the Northern California branch of the ASM introduced a resolution
questioning the advisory committee, its history and exact charge and "the
function of said Committee in regard to the development of techniques of
biological warfare." Before the national meeting, the association archivist
had provided the Northern California branch with documentation from the
1955 meeting of the council that had established the advisory committee.
That documentation identified the committee's vague function as a "liaison"
to the army.[29]

According to the report of the April council meeting published in *ASM
News*, the official society newsletter, the participants approached the issue
from several angles, including the fact that the committee dealt "only" with
scientific aspects of weapons research and that these questions had already
been answered when the committee was originally formed. Even the suggestion that the committee's responsibilities be transferred to the National
Academy of Sciences did not satisfy the representatives from California,
who brought up the issue again at the annual business meeting on May 3.[30]

During the time allotted for miscellaneous and new business, "a motion was entered from the floor 'that the Society discharge with thanks its Committee Advisory to the U.S. Army Biological Laboratories.'" President William Sarles warned those present that "the action of the present meeting would be considered advisory to the council." One unnamed member of the committee seemed to support disbanding it when he "noted the basic unacceptability of biological warfare (BW) as well as other methods of killing people, thus placing BW in a broader perspective of warfare morality." However, he acknowledged "the inevitability and therefore the necessity for acceptance of BW, the potential application of all research on infectious diseases to BW, and the unavoidable necessity that the United States be involved in research on BW." Therefore, he concluded that the committee served a useful purpose, since "microbiologists should have a greater voice in the national effort [and] that secrecy should be reduced to a minimum consistent with national security." Another committee member offered the rationalization that the group "concerned itself only with basic research activity, although necessarily including organisms applicable to BW."[31]

Opinions from the floor addressed many issues surrounding the relationship between an independent scientific organization and a government weapons facility, but no one mentioned the current conflict in Vietnam directly. One individual was concerned that "an official committee tended to lend approval to the activities of Fort Detrick," and since the committee did not have the power to determine what should be done with the research, it "therefore [had] no way of preventing it [BW]." Others asked about the wisdom of giving the ASM name and prestige only to this government office, and were told that other government agencies had not asked for ASM assistance, but that the society would provide it if approached. Before the motion went to a vote, the individual who introduced it made one final comment about the "propriety" of a scientific society engaging in "potentially destructive" work. "The Society, he felt, should neither endorse nor deplore BW research; rather, it should concern itself only with science." The motion was defeated "by a preponderant majority" with only thirty-four supporting votes.[32] The committee was not disbanded.

The next order of society business was the introduction of Salvador Luria as the new president of the ASM. Luria later wrote to a colleague that the 1967 defeat of the resolution "was a big blow at least to my hopes of settling the matter quietly and satisfactorily within the next year."[33] Luria's tenure was not entirely taken up with the Fort Detrick debate, but it was by far the most controversial issue he confronted.[34]

Microbiologists from Northern California were determined to force the ASM to reconsider its role in the development of biological weapons. Alvin

J. Clark of the University of California, Berkeley, wrote Luria a long letter in August 1967 describing his strategy.[35] He tried to organize the various local chapters against the committee and pass a resolution against the committee at a regional meeting. Clark was willing to work on a local and national level to force the ASM to reconsider its relationship with the army. He was frustrated by the arguments put forth at the national meeting that as a scientific society, the ASM should have no involvement in political matters. "I must admit to sympathize with those who would like to have the ASM become more political, however, I am not on the side of those who would have the ASM beat the drums for biological warfare."[36]

Clark encouraged Luria to draft a resolution that would appeal to the scientific sensibilities of the ASM membership. He pointed out how the secrecy of the activities of the committee and the need for security clearance were antithetical to the values of science, and how the support of one particular government violated the tenets of an international scientific society. Clark tried to help Luria identify the "hotheads" in the ASM who were trying to mobilize their peers for another showdown at the national business meeting. He felt that the way to eliminate the committee was by having the CPC change the society's policy, rather than through another contentious debate. Clark pointed out that this disagreement over tactics may have stemmed from the fact that "their desire to do away with the committee stems from their opposition to the war," while he felt that "opposition to the committee should not be based upon the opposition to the war in Vietnam," but rather opposition to weapons development in general.[37]

On November 11, 1967, the Northern California branch hosted a panel on biological warfare and the ASM. President Alvin Clark invited retired Army General Jacquard H. Rothschild, Joshua Lederberg, who was then a member of the faculty at Stanford University and an outspoken opponent of biological weapons, James Moulder, the current chair of the ASM committee, along with Robert Romig, a member of the committee, and Allen Marr, a leader of the North California branch to discuss the nature of biological warfare and the role the advisory committee should have in Fort Detrick's research. The panelists debated the necessity of a civilian scientific role in the military project, as well as the details of what that role should be.[38] There was little agreement among the invited speakers on any of the questions raised by the question of the relationship of the ASM to Fort Detrick.

The debate revolved around two related issues: the specific role of the ASM committee, and the broader question of whether biologists should participate in any weapons research. Moulder implicitly accepted the reality of biological weapons when he encouraged the ASM to maintain the committee, in order to have some say in how microbiology was used. According to

Moulder, the committee "is merely the instrument that is served to remind us of our own personal involvement as scientists, microbiologists, as persons, the involvement of the Society in all the problems arising from the clear possibility of infectious agents being used as weapons." The reminder of the existence of biological weapons was not the issue. "The real problem is what we do about this involvement. This involvement is going to stay with us whether you keep the committee, whether we change it, or whether we do away with it entirely."[39]

In contrast, Lederberg rejected Moulder's reasoning about the need for biologists to have any part of the army research program. "Most of us did not go into science with the expectation of supporting munitions activities. . . . [In general, biologists] had not elected to go into a line of work that would contribute to the destruction of other people."[40] Clark forwarded all of the materials from this meeting to Luria as preparation for a proposal from the Northern California branch at the next national meeting in May 1968.[41]

While the ASM membership debated the future of the committee among themselves, the committee itself was reconsidering its existence at its usual meeting at Fort Detrick in late November 1967. Although they did not make their formal recommendation until March 1968, chairman Moulder communicated their conclusions to Luria in early December. He reported that the committee had "decided to make [two] recommendations to the Society concerning the future," and they hoped that the CPC would "consider . . . [them] as a single possible course of action." The first suggestion was that the committee be "discontinued and that the Technical Director of the U.S. Army Biological Laboratories be encouraged to make more intensive use of nongovernmental scientific consultants than at present." While the ASM could suggest individuals to serve on a panel of advisors, that panel would have no formal relationship to the ASM.[42]

In addition to the reformulation of the relationship between the ASM and the army, the committee proposed that the ASM move headlong into the political arena and "set up a standing committee to be called the 'Committee on Public Affairs,' . . . which would be concerned with the relation of microbiology to matters of public policy, among them . . . biological warfare." They envisioned a new committee composed of individuals with a wide range of opinions, in order to accurately represent the diverse membership of the society. The new committee would not have access to secret documents, and members would not need security clearance. Although this new committee may have trouble convincing military leaders to listen to their opinions, the members of the current committee felt that that representation was necessary. "Our Advisory Committee feels that microbiologists must decide whether they are to have any voice in questions of national and international policy" and that the ASM was an appropriate and "logical" way for them to gain a voice.[43]

In March 1968, the committee made its formal recommendation to the council, and informed Riley Housewright, the technical director at Fort Detrick, of their conclusions. Housewright, a former president of ASM, immediately wrote to Luria to express his "keen regret" over this decision.[44] He was disappointed because of the "high esteem" the Fort Detrick scientists had for the committee, "the many instances over the years of valuable scientific and professional counsel they provided, and the extraordinary vote for continuation of the committee by the membership of the Society" in May 1967. However, the army did not contest the recommendation "in the interest of the harmony that has always characterized the relationship between Fort Detrick and the Society."[45]

At the council meeting on May 5, the committee formally submitted its report and the council accepted the recommendation to discontinue the advisory committee. The committee voted itself out of existence because they did "not believe it . . . [was] serving a real advisory function as presently constituted." Because of its small size and the infrequency of its meetings, the committee was ineffective. Additionally, a recent reorganization of the army unit meant that the committee had virtually no opportunity to influence policy at a high level. "We feel that our Committee (and thus our Society) is now in the uncomfortable position of being expected to furnish technical advice it cannot give and suspected of playing a role in governmental policy it does not have."[46] After the council accepted the decision of the committee, there was a request from the floor to discuss appointing a new advisory committee. Although Luria said he would allow it as new business at a later time, the issue was not raised at the continuation of the council meeting on May 7.[47]

At the ASM's formal dinner on the evening of May 7, Luria delivered his presidential address, entitled "The Microbiologist and His Times." In this speech, later published in *Bacteriological Reviews*, Luria reflected on the responsibility of the microbiologist to his field and to society.[48] This was Luria's opportunity to offer his scientific peers his thoughts on the social and professional responsibilities of scientists. Although he did not directly mention the Vietnam conflict and the uses of science in that war, he used his talk on the role of scientists in American society as a springboard for announcing that the ASM council had agreed to the dissolution of the Fort Detrick advisory committee.

He took the title from an essay by Albert Camus, entitled "The Artist and His Times," which examined the role of the artist and all intellectuals in modern society. Luria noted that Camus, his "favorite writer," raised the issue "of responsibility . . . the recognition of the consequences and implications of one's activities and the willingness to face up to them."[49] He asked his fellow microbiologists to examine two areas of responsibility, to biology as a science, and to society, because in their contemporary world, "it is painfully clear that the findings of science can all too easily be employed, not to enrich the human experience but to make it more painful."[50]

Luria mentioned several of the areas in which microbiology could have serious social consequences, including genetic engineering and eugenics, but he focused his attention on the most obvious example for contemporary society: "the application of science to warfare." Luria was careful to present a balanced view, noting, "There are valid arguments in favor of research on defensive measures against germ warfare." He was honest about the emotional as well as intellectual basis of his own opposition to research on biological weapons, and reiterated his contention that an individual scientist's decision to participate in any kind of war-related research was a personal one.[51]

However, "when the decision was a collective one," Luria felt that the issues were different and simpler. War research, with its accompanying secrecy, was "not fully consonant with the stated purpose of an open-membership scientific organization."[52] He emphasized the fact that the decision to disband the advisory committee was made on technical rather than ideological grounds. Luria said, "It reliev[ed] our Society of a function that in my opinion was not germane to its primary concern with open, unclassified scientific activities." Nevertheless, the dissolution of the advisory committee was also a step toward "the achievement of a society in which science will flourish, both as a liberating intellectual activity and as the source of a beneficial technology." He concluded the speech with an exhortation for each individual member of the ASM to accept the personal responsibility of a scientist, and work so that "society will be so informed and so organized that it can derive the maximum benefit from the fruits of science."[53]

The next day, newspapers carried the news of the dissolution of the committee under the headline "Tie to Army Ended by Biology Society." The short item in the *New York Times* did not cite Luria's address but rather quoted his remarks in an ASM press release, saying that the action had been taken after "the ethical problems implicit in the association of a professional society with the defense establishment" had been discussed by the society, and they had decided to move away from defense work "as a hedge against the use of scientific progress for purposes other than the health and well-being of mankind."[54]

However, the council's decision to accept the committee's recommendation was not yet final, and at the business meeting on May 8, there was "considerable pro and con discussion" about suspending the Fort Detrick relationship.[55] Merrill J. Snyder from the Maryland branch was "shocked" by Luria's address and felt that the press release and the comments attributed to Luria had "distort[ed] the meaning of the Council decision."[56] Snyder later told *Science* magazine, "I don't think that this introduction of the moral issue is in keeping with the views of the membership" of the ASM.[57] He put forth a motion that "the Annual Business Meeting recommends to the CPC and the Council that the Committee Advisory to the U.S. Army Biological Laboratories at Fort Detrick be reinstated." The motion passed with an overwhelming majority, 172 to 58.[58]

On May 9, the officers of the ASM met to try to regain control of the situation. They decided that the issue was significant enough to poll the entire membership of the ASM, in order to have a truly representative decision on an issue "which involves significant matters of the relationship between a scientific society and a governmental institution which carries out a broad spectrum of open and classified research."[59] Luria sent a long memo to all of the members of the CPC and the advisory committee detailing the events of the meeting and asked for comments on the language of the proposed ballot. He apologized if the episode "brought you extra work and concern" but explained how he had tried to avoid this kind of confrontation by using careful language in his speech. Luria had tried to clarify "the distinction between personal activities and Society actions" in his address but took "full responsibility for my statements" if they had been misinterpreted by the press or society members in the aftermath of the announcement.[60] In November, the ASM announced that the entire membership had indeed voted to disband the advisory committee to the U.S. Army Laboratory at Fort Detrick. The people had spoken, and Luria's vision of science in the service of society emerged as the consensus.

The following year, on Veterans Day 1969, President Nixon announced that the United States Army would no longer engage in "offensive biological warfare research."[61] Public sentiment had turned against biological warfare, and Nixon used this opportunity to present himself as a proponent of peace. The laboratories remained open, and in 1972, Nixon designated Fort Detrick one of the first federal Cancer Research Centers under the War on Cancer.[62]

Throughout the ASM debate over Fort Detrick, Luria struggled to keep the issue of Fort Detrick separate from the Vietnam War. However, in comments to the press, Fort Detrick's technical director Housewright expressed

the opinion that the war was directly related to the committee's disband-
ment. He listed "the Vietnam War and the increased concern about the use of
biological and chemical weapons" as well as the fact that "there are those
who say that professional societies shouldn't advise federal agencies." He
implied that the stated desire to keep science away from the government
was merely an excuse, since those same individuals would have no problem advising the NIH on infectious diseases. Bryce Nelson, the *Science* writer
reporting on the issue, argued that the ASM meeting was both "an indication of a shift in attitudes of a portion of the scientific community" and "a
significant reminder that many scientists have not changed their minds
about military-oriented research."[63] Although the issue was closed for the
ASM, the larger scientific debate over biological weapons and other military
ties continued throughout the Vietnam War.

MARCH 4, 1969

As Nelson predicted, "The development of an appropriate relationship to
military research . . . [would] continue to be a subject for soul-searching and
debate."[64] Luria enthusiastically participated in that process of evaluating
the relationship between science and the military in the academic context,
and was a faculty leader in the organization of a campus work stoppage on
March 4, 1969.

Since World War II, MIT had been the top non-industry recipient of
Department of Defense research contracts, with $47 million in defense con-
tracts, along with $80 million going to its federal contract research centers
in the early 1960s.[65] Military research was so pervasive that at times it was
hard to tell whether the campus was "a university with many government
research institutions appended to it or a cluster of government research
laboratories with a very good educational institution attached to it."[66] This
arrangement was beneficial for professors and graduate students alike, since
federal money funded most of their research and provided jobs for scientists outside of academia. However, by the mid-1960s, some of the faculty
and students at MIT had begun to question the dynamics of this relationship in the light of the complex relationship between science and the military, as well as the ongoing war in Vietnam.[67]

Luria was among the first faculty members to join graduate students in
the Science Action Coordinating Committee (SACC) in the fall of 1968.
One of his students, Jon Kabat-Zinn, was a leader of the SACC, and he knew
his advisor would be sympathetic to their cause because of Luria's activity
with BAFGOPI.[68] Luria and Boris Magasanik were the first two members of

the department of biology to sign on, and Gene Brown and Alex Rich eventually joined that winter.[69] Although their department was not dependent on Department of Defense contracts the way the physics and engineering programs were, it still felt the effects of the close relationship with the military that pervaded the institute. The SACC planned the research stoppage on March 4, 1969, in order to spend the day discussing "the misuse of scientific and technical knowledge" by the government and "possible ways for scientists to initiate political action."[70]

Over the course of the fall of 1968, the positions of the faculty and students diverged as the student perspective became more and more radical. The students had conceived of the stoppage as a way to protest the Vietnam War, and indirectly to indict the institute and the entire scientific community for its complicity in the destruction that resulted from it. The faculty members, while sympathetic to the need to draw attention to the role of science in the military, were more cautious in their criticism of the scientific community and focused on national issues as a way to detract attention from MIT.[71] In order for each group to pursue its individual goals, the faculty split off into what became the nucleus of the Union of Concerned Scientists (UCS), which later became an independent scientific public interest organization.[72]

The SACC and the UCS continued to work together on the March 4 program, despite their differences, and eventually forty-eight faculty members signed on as sponsors of the March 4 activities.[73] The two organizations cosponsored the event although each group organized its own sessions. The faculty organizers made it clear that they were in no way striking against MIT. Rather, they emphasized that it was to be a day devoted to "a public discussion of problems and dangers related to the present role of science and technology in the life of our nation."[74] The SACC encouraged similar actions on other campuses, and a dozen other student groups across the country held similar programs on March 4.[75]

"March Fourth Is a Movement, Not a Day" was one of the SACC's slogans, and the event itself spread over three days.[76] During the evening of Monday, March 3, all day on Tuesday the 4th and all day Saturday the 8th, scientists, students, and intellectuals from the MIT community and beyond examined the relationship between academia and the military. Over 1,400 people participated in the events.[77] The Monday and Tuesday programs featured four faculty-run panels that discussed responsibilities of intellectuals, reconversion and nonmilitary research opportunities, the academic community and government, and arms control, disarmament, and national security. The Saturday discussions tended toward larger issues in the social use of science and technology, and included a panel discussion by Luria, Magasanik, and Alex Leaf of Massachusetts General Hospital on "Social Consequences of New Developments in Biology and Medicine," and a panel run by Lewis Mumford on "Applications of Technology to Urban Problems."[78]

Luria signed the faculty statement in support of the work stoppage, worked with the SACC on planning, and chaired the panel on the academic community and the government on March 4, but he did not speak publicly on that day.[79] He only spoke on Saturday, March 8, when he and fellow biologists discussed the "immense" consequences of new developments in biology and medicine, and debated "who is best qualified to make the rules?" Despite the press coverage of the Tuesday events, by Saturday, public interest in the March 4 event had waned, and the Saturday discussions were not reported in the newspapers.[80]

OCTOBER 1969

A few months later, BAFGOPI member Everett Mendelsohn and several other Harvard and MIT professors joined Jerome Grossman of MASSPAX to coordinate a national work stoppage on October 15, 1969.[81] The Vietnam Moratorium was a daylong series of walkouts and rallies all over the country, intended to demonstrate to the government that the American people were tired of the U.S. involvement in Vietnam. The night before, BAFGOPI, along with the Harvard Undergraduate Council and the MIT Undergraduate Association, sponsored a rally "For a Prompt End to the Vietnam War" at Sanders Theater at Harvard.[82] There, Luria introduced the writer I. F. Stone, who shared the stage with economist John Kenneth Galbraith.

On October 15, Luria went to Boston Common to meet with legislators at the State House, attended a peace convocation at MIT, and then joined a protest rally at the village green in his hometown of Lexington, Massachusetts.[83] Zella recalled that he specifically chose to spend Moratorium Day in a location that represented freedom and democracy from the time of the Revolutionary War.[84] He attended the rally with Konrad Bloch, a Harvard biochemist who had fled Nazi Germany around the time Luria left Europe. Luria fondly recalled his emotions on that fall day, when he turned to Bloch and commented, "Aren't you and I lucky, both castoffs from our old countries, to be here today in this unique town, in this wonderful spot, on such an occasion?"[85]

The next day, Luria learned that he had been awarded the Nobel Prize in Physiology or Medicine. He was deluged with letters of congratulation, many from supporters in the Boston area who assumed that he had won the Peace Prize.[86] The pictures of the beaming Luria from his press conference showed a peace pin prominently on his lapel, and Luria promised that part of his prize money would go to support the peace movement.[87]

When President Nixon sent him a telegram congratulating him on the Nobel award, Luria took the opportunity to send his antiwar message directly to the president. He immediately cabled back, "I gratefully acknowledge your kind words of praise and encouragement. I join my voice to that of millions of Americans asking you to stop the fighting in Vietnam and withdraw our armed forces from that devastated country so that we can all work together again for a better America and a better world."[88]

Nixon's congratulatory message took on an added irony, when that Monday, October 20, the *New York Times* and other national newspapers reported that Luria's name was on a list of scientists who had long been blacklisted at the National Institutes of Health (NIH).[89] In June 1969, reporters from *Science* magazine had discovered that the Department of Health, Education, and Welfare (HEW), which housed the NIH, had a secret list of scientists who, for undisclosed reasons, had never been invited to participate is study sections or review panels, although their research projects were still funded by the National Institutes of Health and other federal agencies.[90] Over the weekend of October 17, Luria's now famous name was discovered on that list. Luria issued another statement to the press, expressing his surprise at appearing on the blacklist, since his "personal relations with the National Institutes of Health have been most cordial." He released the text of his congratulatory telegram from Robert Finch, HEW secretary, telling Luria that he had "amply earned the gratitude of all Americans." Luria let this telegram speak for itself, concluding, "I trust that the unwise use of political conformity by N. I. H. and other agencies will promptly be discontinued."[91]

The popular press generally reported this residual McCarthyism in a critical light, but some readers felt that Luria deserved to be on the blacklist for his unpatriotic criticism of the Vietnam War.[92] One Boston area individual shed "no tears for Luria," who did not lose his job at MIT as a result of his political views but was instead "basking in favorable publicity."[93] HEW nevertheless came under fierce criticism from scientific groups. By January 1970, the secretary of HEW announced that they were "eliminating practices" that led to blacklisting.[94]

Another observer saw different political implications for Luria's Nobel Prize. On October 22, Representative Frank Annunzio made a statement recognizing the award in the United States House of Representatives. Annunzio, a prominent Italian American from Illinois, focused on Luria's status as an Italian immigrant (and ignored German-born Delbrück). He enthusiastically proclaimed, "America is proud of her adopted son . . . who came to our shores and by his dedication to his profession, his hard work, and his determination made a great contribution not only to man's never-ending fight to conquer disease but to his adopted country—America."[95] Neglecting to mention exactly when Luria arrived in the United States, he used Luria as evidence that the 1965 overhaul of the Immigration and Naturalization Act, which did away with national origins quotas, "made it possible for our country and our people to benefit from the outstanding contributions of many more great men like Professor Luria."[96]

The ability to sign "Nobel Laureate" on an advertisement or letter to the editor added to the weight of Luria's public protests against the war, as the Nobel Prize conferred additional cultural authority in the public eye. Luria was uncomfortable with this shift in his power, and he was selective about the causes he joined, hoping "that when my political opinion is asked for it is because of my long-standing interest in politics, not because of the decision by a Swedish medical faculty to honor my research."[97] Nevertheless, he continued to accept invitations to speak on behalf of a number of peace groups in Boston, even as BAFGOPI phased out its activities (figure 5.2).[98] In 1970, Women Strike for Peace honored him for his work with BAFGOPI and other public statements against the war.[99]

SCIENCE IN THE SERVICE OF LIFE

By the early 1970s, Luria and some of his scientific colleagues, frustrated by the continuation of the Vietnam War and what they felt was the stubbornness of the Nixon administration, attempted to mobilize their colleagues to commit themselves publicly to the ideal of "Science in the Service of Life." They planned to hold a rally on December 27, 1972, during the annual meeting of the American Association for the Advancement of Science, in Washington, D.C. Luria was joined by two other Nobel laureates who were veterans of BAFGOPI, George Wald and Albert Szent-Györgyi, on a committee organized by historian of science Everett Mendelsohn, who was a vice president of the AAAS at the time.[100]

In the fall of 1972, they wrote to AAAS members, outlining their position and inviting them to sign a letter addressed to President Nixon asking him to stop the bombing immediately. The letter, which was published with only the committee's signatures in the *New York Times* the day after their rally, outlined why they were speaking out particularly during a scientific meeting. They linked the human and environmental destruction in Vietnam

FIGURE 5.2

Luria at an MIT peace protest, circa 1970. Photo courtesy MIT Museum.

when they asked, "Can we as scientists meet in Washington and ignore the fact that our national Administration is launching from this city the most massive air attacks in history? . . . Can we meet to talk of nature as our government is destroying nature?" They reminded their colleagues "as scientists we bear special responsibility." The members of the AAAS must decide, "Is our science to serve life or death?" If the answer was life, then "we must speak out, as Americans, as scientists, against this outrageous misuse of the fruits of science for death and destruction."[101] Mendelsohn announced to the press that they planned to try to deliver the letter directly to the recently reelected President Nixon, although he did not specify how.[102]

The AAAS had already confronted the Vietnam issue since 1969, when radical student groups, led by Scientists and Engineers for Social and Political Action (SESPA), staged rallies and protests at meetings.[103] At the 1972 meeting, the Association administration "took a stand" against SESPA's disruptive behavior, and asked the Washington police to help maintain order. As a result, there were several loud confrontations between protesters and the police, which "overshadowed [the] . . . more sober expression of protest . . . by several prominent scientists distressed at the renewed bombing of North Vietnam."[104] The *New York Times* reported that only about 100 people attended the Science in the Service of Life rally scheduled for noon on the 28th, and that only 250 AAAS members out of the 7,000 attending the meeting had signed the letter to President Nixon.[105]

However, the Committee for Science in the Service of Life had another chance to make their voices heard. At the council meeting on December 30, Mendelsohn successfully introduced "an emergency motion" on the subject of Vietnam that echoed the language of their letter to the president.[106] "In an unprecedented expression of political sentiment," the council adopted a "bluntly phrased resolution" condemning the war. The statement combined scientific and political sensibilities in an unequivocal stance against the current government policy:

> As scientists we cannot remain silent while the richest and most powerful nation of the twentieth century uses the resources of modern science to intervene in the problems of poor and distant lands. Our Association objective, "To increase public understanding and appreciation of the importance and promise of the methods of science in human progress" compels us to refute the view that scientists and engineers are responsible for and endorse, by their actions or by their silent consent, the wanton destruction of man and environment, in this case through warfare.[107]

The committee members were explicit in their acceptance of special responsibility for speaking out against the war by virtue of their unique position in American culture. As scientists and citizens, they had a double charge to reject the violence done in their names. The council also adopted a more specific resolution urging Congress to accept a study by the National Academy of Sciences on the ecological impact of the bombing in Vietnam, which was consistent with an AAAS-sponsored study by Matthew Meselson from 1970.[108]

There was no uniform response from the scientific community to the Vietnam War, just as there was no uniform stance taken by all academics during the conflict. Luria and his like-minded colleagues who tried to appeal to scientific sensibilities to marshal protest represented one part of the wide spectrum of American attitudes about the war. Their position was more visible than others because of their publication tactics, but we have seen that some of Luria's scientific peers were just as resistant to his antiwar message as politicians were. His opponents raised questions of authority, expertise, and representation.

Luria's antipathy to the war extended to the entire Nixon administration. In November 1969, Vice President Spiro T. Agnew ranted against the bias he saw in television news. Luria was so pleased by Agnew's distress that he bought his first "cute little television set" so that he could see for himself the "'distortions' that roused Mr. Agnew's bile so much."[109] In 1973, Brown University offered Luria an honorary degree. When he learned that the U.S. Ambassador to the Philippines, William Sullivan, was also scheduled to receive an honorary degree at commencement, Luria declined the honor. On May 29, 1973, he wrote, "I have made it my policy not to associate in any way with individuals who played leading roles in the planning and waging of the criminal Vietnam war."[110] This policy extended to sharing a dais at university commencements. Luria eventually agreed to accept the degree when he was told that Sullivan would not be attending the ceremony. Later that year, on the day that Secretary of State Henry Kissinger received the Nobel Peace Prize in Oslo, Luria participated in an alternative Nobel Peace Prize ceremony in Boston, presenting "a symbolic certificate on behalf of the Vietnamese and American peoples" that noted that more than 100,000 people had been killed or wounded since the cease-fire that Kissinger had brokered.[111]

Throughout the Vietnam era, the FBI carefully monitored Luria's antiwar activities and kept a record of when Luria participated in teach-ins or organized a newspaper ad during the Vietnam War.[112] Other individuals contacted the FBI with their suspicions about Luria's loyalty. In 1969, a Boston area citizen, identified only as "JM" in Luria's FBI file, sent a letter to J. Edgar Hoover, reporting "what looks like a cabal or communist plant on nearby campuses, centering upon three faculty members," including Luria. These suspicious professors "apparently work together . . . in persistent trouble-making, rabble-rousing, and in anti-war and anti-ROTC agitation." The writer suggested to Hoover that they may "be worth looking into or they may simply be half-baked nitwits (but certainly far from harmless)."[113] Hoover thanked the writer for his "thoughtful" letter and sent him some materials about campus peace organizations. After Luria spoke out against Nixon's reelection, one Boston citizen was so upset by his remarks that he wrote directly to the president about it.

I wish you would have the F.B.I. investigate this Proffessor [*sic*] Salvatore [*sic*] Luria from Harvard. In a speach [*sic*], he accused you of being a fashist [*sic*] a Hitler and a Stalin and suggested you Mr. President should be impeached. I don't believe he is an American citizen. I am sure the F.B.I. will find him an Italian Communist and should be deported at once.

The News Media has been broadcasting his remarks for the last 36 hours, on the hour . . . I am sure Mr. President we don't need or want the likes of that in this country, since we have enough of our own American born.[114]

Although the 1960s was a period of questioning all authority, including that of science, Luria was still able to use the rhetoric and moral weight of science to lend legitimacy and authority to his political position.[115] In this period it was impossible for Luria to separate his scientific and political commitments in his thoughts and actions. In the political arena, he was able to raise money and publish ads because he had extensive ties in the scientific community, and his message was broadcast and his authority was accepted, at least by some, because of his stature as a scientist. In the scientific community, Luria would not let himself or his peers isolate themselves by rationalizing that their work was "pure" and thus avoid the uncomfortable possibility that their work had somehow contributed to the war. Luria integrated his political and scientific ideals into a single passionate determination to use the ideals of American democracy and scientific integrity to work toward a positive future for humanity. His dedication to protecting American democracy from a destructive war and his commitment to ensuring that the work of science would always be used for the benefit of mankind compelled him to oppose the Vietnam War. As an American, a scientist, and a human being, he felt he could do nothing else.

6 BIOLOGY FOR AMERICAN SOCIETY

THE NATIONAL CANCER ACT OF 1971

In the late 1960s, Mary Lasker and other lay leaders of the American Cancer Society felt that government funding for cancer research was inadequate. The Citizens Committee for the Conquest of Cancer ran a campaign to direct public and legislative attention to substantially increasing the budget for cancer research. Comparing cancer to the moon shot, they argued optimistically that if only enough money were thrown at the problem, cancer could be cured by the Bicentennial in 1976. The 91st and 92nd Congresses heard evidence, received recommendations, passed proposals, and debated the merits of focusing national research dollars on the conquest of cancer. In December 1971, President Richard Nixon declared war on cancer, and signed into law the National Cancer Act. The National Cancer Institute was expanded and partially removed from the National Institutes of Health while millions of dollars were allocated in the federal budget.[1] During the year and a half of political debate over the War on Cancer, Luria was one of the many scientists who were called before the Senate and House committees on health to testify on the feasibility of such a project.

Lasker and other representatives of the American Cancer Society were concerned that in the wake of severe National Institutes of Health (NIH) budget cuts in 1969, funding for cancer research in particular would plummet, and they lobbied for more support. In response, Senator Ralph Yarborough of Texas introduced a bill calling for renewed effort in cancer research in March 1970. The entire scientific community was concerned about the tightening of the NIH budget, but Lasker and her fellow "benevolent plotters" brought the cancer issue to national and legislative attention.[2] There was no reason, Yarborough and other senators thought, that the same scientific initiative that Americans had shown with the moon shot could not be put to good use in the fight against cancer. A panel of experts was appointed to report to the Senate on the current state of cancer research and make recommendations for a new cancer institute, which would be independent of the National Cancer Institute and whose director would report directly to the president.[3] Both the president and the Congress were eager to support the idea that "the conquest of cancer is a national crusade."[4] Although Yarborough was voted out of office in November of that year, his cancer initiative continued.

In the spring of 1971, the Senate debated both their version of the Conquest of Cancer Bill, S. 34, and the president's version, S. 1828, to "amend . . . the Public Health Service Act to provide for the conquest of cancer" by creating a National Cancer Authority to subsume the National Cancer Institute and operate outside of the NIH.[5] Virus research was prominent in the plan, since many scientists agreed that it was only a matter of time before the elusive human cancer virus would be found. Senator Edward Kennedy was the new chairman of the subcommittee on health, and he and veteran Republican senator Jacob Javits guided the bill through hearings and debates throughout the spring 1971 Senate session.

Many scientists, including several who served on the Senate advisory panel, expressed profound concerns that the bill would undermine rather than encourage cancer research. While they shared the desire "to realize scientific and clinical advance in the war against cancer," scientists doubted the viability of the legislative agenda.[6] The narrow focus on directed research, they felt, was not a logical way to approach the problem, and the budgetary isolation of the proposed agency would put the rest of the research sponsored by the National Institutes of Health in jeopardy. The scientific community overwhelmingly felt that "everyone would lose if cancer were removed from the rest of biomedical research: both cancer . . . and the broader biomedical research effort would be impoverished."[7] Nevertheless, on July 7, 1971, the Senate voted to approve S. 34, with its provisions for a separate agency, by a count of 79–1.[8]

The bill immediately moved to the House Subcommittee on Public Health and the Environment of the Committee on Interstate and Foreign Commerce. Florida representative Paul Rogers chaired the subcommittee, and he shared the scientists' skepticism over the need for a new Cancer Authority.[9] Rogers introduced a substantially different bill, H.R. 19681, which still called for additional funding of cancer research, but the new authority would remain under the NIH umbrella, and would be part of a broad medical research agenda. After he introduced his version of the "National Cancer Attack Amendments," Rogers scheduled several rounds of hearings for the fall of 1971, in order to ensure that the administrative and scientific concerns would be addressed.[10]

On Tuesday, September 28, 1971, Luria appeared before the subcommittee as a representative of the American Society for Microbiology, but by extension, he spoke for many other experimental (as opposed to clinical) researchers.[11] He did not support the Senate bill. Despite his public opposition to many of President Nixon's policies, Luria reassured the committee that his "concern [about the proposed law] is strictly scientific" as he commented on the structure of the proposed agency, its funding, and the effect it would have on training programs in the United States.[12] After reading a prepared statement, he answered questions from several members of the subcommittee, bantering and joking with some of them, and always speaking from a scientific, rather than political, perspective.

Although the American Cancer Society's Committee on Growth had supported research, including his own, that made great strides in the understanding of basic life processes, Luria felt that the scientific community was not "ready for a frontal attack on cancer, in the same way as in 1950 we were ready for a frontal attack on poliomyelitis and in 1960 for a frontal attack on space travel."[13] Those endeavors, he pointed out, dealt with technical problems, but cancer research was not at the point that the only problems that remained were technical. Given the current state of cancer research, he felt "that we can work out an effective, expanded, rationally, and soberly conceived research program," but it would be unfair to expect rapid results. "Any vision of a crash program promising a 'cancer cure' in 3 or 5 or 10 years would be a self-delusion and a dangerous misleading of the public."[14]

Nevertheless, Luria was fully committed to increasing federal funding for basic research, and keeping the cancer initiative within the auspices of the NIH as Rogers had proposed in his bill. Like other scientists, he was concerned that the proposed Cancer Authority was "the first step in the dismantling of NIH."[15] He also objected to the possibility that the new authority would favor research contracts over grant research, since contracts had the potential to undermine the peer review process that distributed research grants and could also push basic research to the sidelines. Finally, he expressed concern about the lack of graduate and postgraduate funding in this bill in particular and in the science budget in general. He reminded the committee that reducing fellowships for graduate students "is the wrong direction to take if we want to succeed in what we set out to do in science—be it going to the moon or conquering cancer."[16]

Committee members pressed Luria on several of his points. Representative Nicholas Kyros reminded him that some of the rationale behind removing the new Cancer Authority was to expand the agenda beyond the research focus of the NIH to include cancer cures as well. Luria replied that he had no problem with that focus but he wanted to make sure that "the integrity and strength of the NIH" be maintained, even if it had to expand to test possible cancer cures as well as support basic research. He was more concerned about "this hullabaloo with the Director reporting to the President—which to me is poppycock." He continued jokingly, "Not that I don't expect the President of the United States to have the last word on everything, but if I had to get treatment for cancer, I would prefer to go to a physician."[17] Representative Alcee Hastings thanked him for his comments about the rhetoric of cancer research, since "people in our position tend to think that the passage of a bill is going to provide all of the magic solutions."[18] Other committee members asked for clarification about the role that viruses might play in tumor development, but no one challenged his assertions about the need to protect the political and scientific independence of the NIH.

After his testimony, Luria continued to communicate with lawmakers as they debated the risks and benefits of the proposed research structure and as the White House and Congress traded versions of the bill. In late November, just before the final votes in the House and Senate on the National Cancer Act, Luria wrote to Senators Kennedy and Javits, reminding them of the "importance of maintaining the efficiency and morale of this unique organization NIH by protecting it from political influences." He encouraged them to consider all of the ramifications of the proposed legislation, since it "may accelerate the conquest of cancer but may also, if unwisely structured, cause irreparable damage to the greatest institution for medical research in the world."[19]

President Nixon declared War on Cancer when he signed Public Law 92–218, the National Cancer Act, into law on December 23, 1971. This "wonderful Christmas present" to the American people amended the Public Health Service Act "so as to strengthen the National Cancer Institute of Health in order to more effectively carry out the national effort against cancer."[20] In a concession to the scientific community, Rogers had insisted on maintaining the integrity and authority of the NIH, and the National Cancer Program remained a part of the National Cancer Institute.[21] However, the act stipulated that the new cancer program would submit its budget requests independently of the NIH, and that a National Cancer Advisory Board, which was appointed by the president, would administer it.[22]

THE MIT CENTER FOR CANCER RESEARCH

The National Cancer Act provided for the establishment of fifteen new cancer research centers at academic institutions around the country, to focus on
clinical and basic research.[23] Luria and his colleagues at MIT saw this as
an opportunity to expand their research program in virology and genetics
and to strengthen the place of MIT in the national life sciences community.
David Baltimore had recently discovered reverse transcriptase as part of his
RNA tumor virus project, which was hailed by the scientific and lay press
as a breakthrough in the understanding of tumor development.[24] The move
to a concentrated cancer research agenda at MIT seemed to be the next logical step. Although it may have seemed unduly opportunistic in the light of
Luria's criticisms of the National Cancer Act, the founding of MIT's Center
for Cancer Research was not a feeble attempt to get a few more dollars for
basic research in exchange for lip service to the problem of cancer. The definition and goals of the center were consistent with Luria's vision of how to
approach cancer research, as well as with the scientific culture at MIT.

In anticipation of the available funding, Luria began drafting a proposal
for "A Program in Cancer Cell Biology" in the summer of 1971. He had been
named an Institute Professor in 1970, which lightened his teaching load,
but he agreed to take on more administrative responsibility as the director
of a proposed Center for Cancer Research. Attentive to the political climate
for biomedical research, Luria tailored the grant application to combine the
needs of the MIT department of biology with those of the larger national
research establishment. Early in 1972, he submitted the grant proposal to
the National Cancer Institute.

His research proposal echoed his congressional testimony when he noted,
"The cancer problem is not ready for a short-term frontal attack of the crash-
program type."[25] Although new developments unfolded regularly, "it would
be naïve . . . to consider the battle against cancer a short range project."[26]
Given the complexity of cancer biology and the "multitude" of types of
cancer, any research program had to include a graduate training component
in order to ensure a steady supply of research manpower.

Luria felt that MIT met the criteria for a "favorable situation" for a successful program in cancer pathobiology. The institute was already an interdisciplinary environment with the potential to "establish ready contract
with clinical facilities for cancer patients within strong general hospitals
and . . . attract first-class people for training in cancer research."[27] Luria
argued that the institutional culture of MIT made it the ideal setting for a
Center for Cancer Research. Current MIT faculty and resources in biology,

biochemistry, chemistry, physics, engineering, computer science, and communication science would continue to complement the institute's "close and cordial" relations with Boston area teaching hospitals.[28]

The research proposal for the new center focused on four broad but related areas in cancer etiology and pathology on a cellular level: virology, cellular biology, developmental biology, and immunology. All four seemed to be promising avenues of research, and several MIT department of biology faculty members were already working in those areas. Each of these areas presented the opportunity for interdisciplinary collaboration, and Luria anticipated that the relationships between the various groups would be "intimate and continuous." In addition to the four research foci, Luria asked for resources for complementary chemical studies and for setting up a communications network for cancer researchers around the world.[29]

Using recent literature, Luria presented convincing evidence that the four research programs represented the cutting edge of cancer research. Research on tumor viruses was an especially promising field, from studies of RNA and DNA infection, metabolism, and infection to viral antigens and mutations.[30] RNA tumor viruses in particular seemed to be a good choice, especially because they seemed to be "the most likely candidates to be the cause of at least some human cancers."[31] David Baltimore was named an American Cancer Society Professor of Microbiology, which brought additional research funds to the cancer center, and ensured that the MIT team would make significant contributions to research in tumor viruses.[32]

MIT's department of biology would benefit substantially from the new facility. The Center for Cancer Research would allow the department of biology to expand beyond its quarters in the Center for the Life Sciences in Building 68 on Ames Street. The institute was willing to offer three floors in a building nearby that was formerly a chocolate factory, with 69,000 square feet of usable space. Additionally, the department of biology would gain faculty members, since researchers would hold appointments in both the center and in their areas of expertise.[33]

Luria's proposal met the NCI's standards for new cancer centers, and the NCI pledged support for renovations of the existing building as well as for operating costs for the first four years. On December 4, 1972, the MIT administration announced the federal grant. In addition to the $2,362,500 that the NCI promised for the physical construction of the center, the MIT Corporation offered $1.8 million. The center was scheduled to be in full operation by 1975.[34]

As director, Luria was involved in all aspects of the new center, from the plumbing to the laboratory equipment to the interior decoration. He was "extraordinarily pleased" with the results. He later commented, "I do not remember enjoying any single task in my life more than designing and supervising the remodeling of that building."[35] In addition to such mundane tasks, he threw himself into recruiting the top researchers in immunology, cell biology, virology, and developmental biology. In the summer of 1973, Luria focused on the task of hiring six new faculty members, and turned to some of his old colleagues, including Delbrück and Watson, for assistance.[36]

Luria identified two promising candidates who were completing post-doctoral fellowships under Watson, who had been named the director of the Cold Spring Harbor Laboratory in 1968.[37] Watson was also interested in the relationship between viruses and cancer genetics, and one of his first acts as director was to establish a DNA tumor virus lab.[38] Nancy Hopkins was "thrilled at the opportunity" to join the center in 1973, and Phillip Sharp followed in 1974.[39] Although Luria "protected his young faculty from unnecessary interruptions," nearly all of the staff took on a full teaching load in the department of biology.[40]

The approach of combining basic research projects in virology with the cancer agenda in the context of molecular biology yielded fruitful results within a few years of the establishment of the Center for Cancer Research. In 1977, Phillip Sharp discovered messenger RNA splicing, for which he won the Nobel Prize in 1993.[41] Viral oncology research done by Robert Weinberg at the center in the early 1980s played a role in the isolation of oncogenes, mutated genes that promote cancer growth.[42] While oncogenes opened up an important new area of genetic and cancer research, their discovery and elucidation effectively stopped any new research into the viral causes of human cancers. Hopkins's research on RNA tumor viruses provided valuable insights into leukemia viruses before she turned her attention to developmental genetics, and she had remarkable success using zebrafish as a productive model organism.[43]

In light of these and other scientific successes, Luria's contention that the MIT center comprised "one of the best outfits in basic cancer research" seems to have an extra word in it.[44] In the years since the National Cancer Act invigorated the biomedical research community with extra dollars and resources, "cancer research" and "basic research" have become synonymous. Projects on a wide range of biological questions about DNA and gene action are often funded through cancer programs or take place in cancer research laboratories but more often than not provide new insights into areas of biology that have no direct bearing on cancer etiology, pathology, or treatment.[45] However, in the political and social climate that biologists inhabited in the aftermath of the legislation of 1971, basic biology research came to be defined as somehow relevant to the War on Cancer. This orientation fit in well with the culture of MIT, where the divide between basic and applied research was often blurred, and resonated with the American political expectation that scientific research have a practical component.

POPULAR SCIENCE

On October 20, 1969, just after he won the Nobel Prize, Luria published his first essay on biology and society for the general public in the liberal magazine *The Nation*. The article was entitled "Modern Biology: A Terrifying Power," but the focus of the piece was not on the dangers inherent in the new science but rather on how new ideas were being used or ignored. The misleading title came from the editors of *The Nation*, not from Luria, who was unhappy about "alarmist titles from magazine editors."[46] The essay combined a short lesson in biology with a call for more communication and social involvement by scientists along with a critique of antirational trends in American society and politics. This was the consistent underlying theme of most of his public comments about science over the next twenty years.

Luria emphasized the need for "a rational decision-making machinery" that would advise national and international bodies on how to apply emerging scientific knowledge. He argued that it was a political, rather than scientific, task "to create a society in which technology is purposefully directed toward socially chosen goals." Luria felt that this kind of rational advisory committee could have prevented such wasteful projects as the "man-on-the-moon venture" and help an otherwise "organized society cope with the menace of overpopulation, a threat greater than that of nuclear self-destruction."[47] Although he did not give specific suggestions for establishing the rational decision-making machinery, Luria did present a list of things that scientists could do to "face the problems within their own sphere of activity." They should be the first ones to examine the social implications of their work and be willing "to tell society, in a forceful and persistent manner, what science is discovering and what the technological consequences are likely to be." He concludes his wish list with a plea for scientists to take on more leadership roles, in order "to awaken the public and their elected representatives from the complacency that lies behind the distorted priorities of present-day society."[48]

In other messages to his fellow scientists, he was more willing to draw close connections between science and a utopian world. For example, in the 1970 Hermann J. Muller Memorial Lecture, Luria charged "We need to restore [a] . . . vision of a humanity rationally planning its own future—using the fruits of science to substitute happiness for suffering, cooperation for competition, common effort for war." Scientists not only had to provide new knowledge and participate in the decision-making process, they also had to resist the impulse to take themselves too seriously. "We must not delude ourselves as to the nature of our own contributions." Free will on the part of every citizen was the key to a democratic system, and all scientists could do was ensure that as many citizens as possible rationally exercised their free will.[49]

LIFE: THE UNFINISHED EXPERIMENT

In 1969, Luria's friend Theodosius Dobzhansky invited him to join in a publishing venture for Scribner's Sons, writing a pair of books that "would present to an intelligent but non-specialist reader what is meaningful and exciting about our science."[50] They agreed that Luria would write one book on molecular biology, and Dobzhansky would be responsible for the volume on organismic biology, under the editorial direction of Kenneth Heuer at Scribner's. Luria immediately put together an outline of a book and began writing.[51] Dobzhansky did not deliver his proposed volume because he was distracted with other projects, but in 1973, Luria published *Life: The Unfinished Experiment*.[52] Some editions of the 150-page work had the subtitle "A Nobel Laureate Interprets Modern Biology." The title was originally going to be *Life: The Unique Experiment* but when Nancy Ahlquist typed up the manuscript, she read his handwriting as "Unfinished." He liked that title better.[53]

In the book, Luria outlined his broad vision of biology. He began with the idea of evolution, and the role natural selection has played in shaping all life, from bacteria to man. He emphasized the nondirectionality of evolution, reminding his readers that genes, species, and nature do not act in any conscious way: "Man is alone in knowing the joys and torments of conscious will."[54] In order to explain what he called "the central problem of biology—the nature and function of the hereditary material and its relation to evolution," Luria detailed how the hereditary material acts on a genetic, cellular, and species level.[55] He showed how evolutionary pressure explains how living things use energy, how their bodies are organized, and even the complexity of higher organisms and the relationship between those organisms and the environment.

Luria devoted one chapter to the evolutionary development of man, again emphasizing that modern humans, with their knowledge of and power over nature, do not represent the culmination or a goal of nature. This chapter includes a discussion of Luria's perspective on some of the pressing biological issues of the day, including overpopulation, genetic engineering, and abortion. He concluded the book with some suggestions about how the human mind may have evolved, and what the evolutionary, biological purpose of human intelligence may be. The book ends on a celebratory note, with Luria attributing to natural selection those human mental characteristics that represent "the innermost sources of optimism—art, and joy, and hope, confidence in the powers of the mind, concern for his fellow men, and pride in the pursuit of the unique human adventure."[56]

Although Luria often referred to contemporary research and uses technical language, the book does not assume any scientific knowledge on the part of the reader. The prose is straightforward and concise, and the book has only two illustrations, of how DNA molecules work. He provided an extensive glossary, and a list of college and high school level textbooks for further reading.

Luria's political and social priorities framed the way he presented biology to the public in this popular and well-received book. Through his language and choice of imagery used to describe scientific ideas, Luria included a fair amount of social criticism and subtly argued that science could also provide a model for human society, at a time when political and social institutions were in flux. His suggestions for science-based social change reflected his political perspective on some of the key debates of the 1970s, including feminism, birth control as a way to prevent overpopulation, and eugenic abortion.

In some cases, Luria was explicit about the ways in which scientific literacy could change society. In the introduction, he noted that he tried to avoid using "deplorable" sexist language such as "man" and "his" when referring to the human species as a whole, but was "bogged down in the clumsiness of alternative expressions," and so he apologetically used the term "man" with full acknowledgement that men and women are represented roughly equally in the species. Science books such as this one could potentially change the sexist conditions of society, though, since "an understanding of biological interdependence among all members of a species may help foster the recognition of the justness of social equality between members of the two sexes and among all groups of mankind."[57]

A scientific view of life and humanity also provided Luria with the opportunity to argue for the uniqueness and intrinsic value of each individual—a stance that led him to lament the racism that was still rampant in American society. When discussing the complexity of cells and organisms, he marveled at the immense variation of genes, made possible by the "infinite nuances of heredity." Luria dispelled any scientific justification for identifying or judging groups according to race and celebrated the fact that "Not only in his thoughts, his feelings and his will, but in the chemical markings of his body each human individual is unlike any other that has ever existed."[58] Luria's feminist sensibilities had developed slowly over his time in the United States, and by the early 1970s he was fully committed to equality in the home, in the workplace, and in society, and felt that scientific data would support that feminist stance.[59]

Luria took great pains to emphasize the ways in which "man is but one product, albeit a very special one, of a series of blind chances and harsh necessities."[60] Humanity is "exceptional, unique, and troublesome," in the ways in which it manipulates and modifies the environment. The chapter on man reviews human technological and scientific accomplishments from fire and agriculture to modern medicine. The effects of this progress are problematic, not only because of environmental destruction, but also because of the threat of overpopulation.[61] Luria acknowledged that the problem is at once scientific, social and political, but ultimately cast his vote with those sociologists and biologists who "believe that the successful achievement of a humane regulation and limitation of human population growth is the most urgent prerequisite to attaining harmony with the world and to perpetuating an acceptable human life."[62]

Luria presented natural systems as models for human social organization. After a lengthy description of fermentation and respiration in the chapter entitled "Energy," he concluded with a discussion of how natural selection maintains all efficient and effective ways of using energy. He then made a rhetorical leap to human society. Luria observed, "In evolution as in human affairs, surely the greatest wisdom lies in preserving a balance of mutually reinforcing, mutually complementing ways of performing a task."[63] In another section, Luria argued that it would be possible to construct an evolutionary scenario in which natural selection selected for "cooperative behavior that favored communal living."[64] In addition to justifying social equality, science could be an inspiration for cooperation and efficiency.

Luria also took the opportunity to dispel sociobiological explanations for human behavior while offering pointed social criticism.[65] He claimed that those who would draw analogies between animal and human aggression were making "a scientifically unwarranted jump" in order to justify "war, criminality and racial strife" as biological urges that are based on evolutionarily advantageous traits. In Luria's understanding of the biological and sociological data, "aggression in human society is due much less to biological imperatives than to sociological imperialism—that is, to the organization of society itself."[66] Natural selection could some day exert genetic pressure on human behavior, but "only if man's culture, utterly misapplied, should create extreme stress situations such as excessive overpopulation or intolerable pollution or famine."[67]

Science magazines such as the *Quarterly Review of Biology* and *Scientific American* called the volume "lucid and authoritative," and praised Luria for writing that was at once "humane, sensitive, . . . genuinely clear, . . . literary in the best sense; the flowing text is personal, informed [and] honestly meant for the general reader."[68] Edward Edelson, the science editor of the *New York Daily News*, identified Luria as one of the "elder statesmen of science" when he discussed the book in the pages of *Smithsonian*. Edelson admired Luria's use of biological facts, and his success at achieving the goal of explaining biology and evolution in a way that made the book "essential reading for almost anyone."[69] In addition, several of Luria's scientific colleagues, such as Dobzhansky and René Dubos, wrote to him praising the book and offering encouragement and discussing different scientific and philosophical points.[70]

With titles such as "Gene Drives a Hard Biological Bargain," "Intricate Beauty of Our Molecules," or simply, "Biology for the Layman," reviews from newspapers in Cleveland, St. Louis, Los Angeles, New York, and Boston demonstrate that Luria achieved his educational goal.[71] The reviewers agreed that Luria's prose and explanations were among the clearest they had seen. The *Cleveland Press* and the *St. Louis Post-Dispatch* mentioned that the book would be an important resource for ordinary citizens who would have to make decisions about "the uses of biological knowledge in human affairs."[72] In the *New York Times Book Review*, Gunther Stent emphasized Luria's "urbaneness, modesty, reasonableness, and lack of cranky righteousness."[73] Kenneth Klivington in the *Los Angeles Times* noted, "It's not a spy thriller. But . . . it's the best source to date on what the DNA hubbub is all about."[74] Klivington criticized Luria's characterizations of man and his mind, and disagreed with many of the liberal social programs Luria suggested.[75] He also expressed some broad skepticism of the "revolutionary" promises of molecular biology. Similarly, Herbert Kenny, the *Boston Globe* reviewer, cautioned that Luria's book was "not easy reading, although Professor Luria, a man of charm, wit and lucid language—at times almost poetic—has put the more complex portions of his presentation about as simply as they can be put." Kenny noticed that Luria did not include faith on his list of the sources of human optimism, an observation that Luria himself made in a letter to René Dubos, when he described *Life* as a "modestly but honestly an anti-religious book."[76]

On April 18, 1974, Luria was awarded the National Book Award in the Sciences at a memorable ceremony at Lincoln Center in New York City. Luria's fellow awardees included the poets Allen Ginsberg and Adrienne

Rich, and novelists Thomas Pynchon and Isaac Bashevis Singer. The ceremony was marked by the unexpected arrival of a streaker, who ran through though the hall naked, shouting, "Read books! Read books!"[77] Reports of the ceremony included praise for Luria's wit and comic flair. "Bounding to the microphone, [Luria] cried in his inimitable accent: 'No check, no speech!' and then, having received the money, launched into the drollest acceptance of the evening."[78]

Luria was "absolutely delighted" by the award and commented that it was more enjoyable than receiving the Nobel Prize because there was "no royalty—plenty of royalties." Although there was no way he could have known about the streaker in advance, Luria must have elicited quite a laugh when he told the audience that he had tried to "send science streaking along in its glorious nakedness, stripped of the veil of professional jargon and without even a plastic raincoat of pseudo-philosophy." He mentioned that it was his and Zella's twenty-ninth wedding anniversary, and he acknowledged her role as "the companion of my own unfinished experiment."[79] In an analysis of twenty-five years of National Book Awards ceremonies in the *New York Times Book Review*, poet and author William Cole praised Luria for being "as witty as anyone I've ever heard on a platform," and described how he delighted in Luria's appearance, who "in approach and accent was for all the world like my favorite Sesame Street character, the Count."[80]

SOCIAL JUSTICE BEGINS AT HOME

Around the same time, Luria sat down to write up his thoughts on an "unusual experiment" he and Zella had undertaken in sharing household responsibilities.[81] He had long prided himself on being a "helping husband," and was amused by the media focus on the fact that he had been washing the breakfast dishes when he heard he had won the Nobel Prize.[82] But one evening over dinner with other two career couples, Luria realized that full responsibility for planning and cooking meals "was every bit as much a demanding an absorbing occupation as was for me the teaching of students and the performing of experiments," and that Zella and the other wives "had actually two jobs, carrying a dual load of responsibilities, as professional worker and homemaker."[83] To his surprise, Zella agreed when he proposed that they share those tasks equally. Each month, they would alternate who would be responsible for shopping, cooking, and cleaning up. The other partner would be free to relax in the living room with a pre-dinner drink.

One month into the experiment, Zella was entertaining an Israeli guest Luria had brought home. While he was cooking, Luria realized, "My God, for twenty-six years it has been I who sat with guests, sipping my drink while she was cooking! It was then that the meaning of inequality hit me with full force."[84] Once he acknowledged the inequality in his own relationship, he appreciated the feminist slogan, "The personal is political." He began noticing the assumptions and expectations about women's roles and domestic arrangements that were still prevalent in the 1970s workforce, and tried to dismantle them in his own professional settings. Over the next few years, Luria learned to read cookbooks, ask his hostesses for recipes, and develop his own signature recipe for "Scrambled Eggs Special."[85] More fundamentally, he also came to realize that sharing these household duties was "an exercise in internalizing the idea of equality between man and woman . . . [an] indispensable step toward becoming a fully integrated, just human being." He understood that "the unquestioned assumption of male superiority . . . serves as the backbone for all sort of prejudices . . . of race, creed, or nationality."[86] Every soufflé he made became "a small contribution to social justice."[87]

FIGURE 6.1
Salvador Luria cooking at home in Lexington, 1970s. Photo courtesy of Daniel D. Luria.

WHAT CAN BIOLOGISTS SOLVE?

Throughout the 1970s, Luria argued that scientific rationality was the key to maintaining democratic systems and humanitarian social policy. However, he also argued that scientists should not be given primary responsibility for problems that, while they may have biological or medical components, are inherently social or political. In a controversial essay published in the *New York Review of Books*, he described how biologists were "being pressed to take on" three issues that were "not, or at least not primarily, biological problems, but [rather] social problems" with biological components.[88] Ecological disaster, violent crime, and racial disparities in IQ scores all pointed to the need for "radical changes in social priorities and improved machinery to enforce those priorities." While there was a role for applied biology in defining and addressing each of these problems, Luria was worried about identifying them as scientific problems that only scientific expertise could solve. "If scientists are lured into claiming that they have the know-how to solve what are really social crises, they will share the responsibility for the fact that these crises remain unsolved."[89]

Luria was particularly concerned about the recent public discussions about whether violence and intelligence are innate and fixed human qualities. He argued that "biologizing" violent crime as part of human nature only "serves to make people close their eyes to what crime really is: a social illness fed by poverty and by profit," a by-product of social exploitation in an industrial society.[90] Similarly, the use of "shaky and probably meaningless evidence" by psychologists and educators who were arguing that the difference between Black and white students' scores on standardized intelligence tests could be explained by genetic differences between populations opened biologists and social scientists alike to charges of supporting "racist eugenics."[91] Luria referred to several recent studies that demonstrated that economic opportunities, which often correlate with white skin, were a more valid indicator of future success than IQ scores. He warned against stumbling into the "quicksands of the genetics of IQ," since biologists could well "end up as the stooges of the forces of racial bigotry."[92] This call for caution did not contradict his earlier calls for scientists to accept their social responsibilities. Rather, it reinforced his charge to biologists to commit to educating the public about scientific reasoning in general and the uses of genetic knowledge in particular. Falling into "socio-political traps beyond the scope of science" would not be the best way for scientists to fulfill their responsibility to society.[93]

Not surprisingly, Luria's comments about race and IQ generated a strong response. Stanley Bosworth, the headmaster of Saint Ann's Episcopal School in Brooklyn, wrote an impassioned letter to the editors of the *New York Review of Books* criticizing Luria's discussion of the predictive value of IQ scores.[94] Bosworth gave a short, positive review of the uses of intelligence testing, and accused "Mr. Luria" of "Lysenkoism" for suggesting that scientists should not research the genetics of IQ.[95] He resented the implication that "educators and psychologists concerned with IQ are bigots, stooges and racists" and felt that Luria's suggestions for how scientists could contribute to expanding educational opportunities amounted to "medieval proscriptions of research and inconvenient facts."[96]

Luria's published response reminded Bosworth and the readers of the *New York Review of Books* that intellectual traits are not the same thing as IQ, and that the measurement of intelligence is fraught with difficulty. He assured his audience that he does not consider those involved with IQ research stooges, bigots, or racists, but was concerned that "less sophisticated" researchers might unwittingly become the pawns of such individuals. Finally, he gave a strong defense of critical treatment of scientific methodology and research programs, saying that it is the process of mutual criticism that "keeps research reasonably honest and meaningful."[97]

The anthropologist Ashley Montagu sent the editors a long letter that began with a strong agreement with Luria's position on IQ testing. Much of Montagu's career had been taken up with confronting racism in science, and his classic 1942 book *Man's Most Dangerous Myth: The Fallacy of Race* was considered one of the most effective tools in combating scientific explanations of racial superiority.[98] He discussed the historical context of IQ tests, and made the strong claim that "both terms, 'race' and 'IQ' are fraudulent" in that they hide social and economic interests under the veneer of scientific objectivity. Montagu attacked the logic of IQ testing, arguing that the tautology "'intelligence is what intelligence tests measure' is scientifically absurd and socially irresponsible." He argued instead that intelligence is "a function of highly complex variables" and that environmental factors cannot be meaningfully separated from genetic ones.[99]

Luria was "grateful to Professor Montagu for clarifying further the IQ issue" but pointed out that he differed from Montagu in his challenge to the "integrity of purpose behind the IQ issue." He felt that the IQ issue was a smoke screen for assumptions about race, not intelligence. Luria bluntly stated, "If those who deal in race and IQ know what to do about improving IQ, or improving schools for low IQ students, let them do it for white and black alike. If not, let them stop peddling mischief or be exposed for what they are—racists."[100] Even if his critique was ignored or dismissed, he would not stand by and let fellow scientists—and fellow Americans—use scientific research to promote agendas that perpetuated inequality and discrimination.

THE HUMAN GENOME PROJECT

Luria was prominent enough to be featured in a number of news reports on the impact of immigrants on American society in the early 1980s. He was sometimes reluctant to participate, as he did not think of himself as Italian American—just as an American.[101] He also harbored some resentment toward the Italian government. In response to an invitation to speak at a Nobel Lecture series at the Italian consulate in 1986, he wrote,

> Your letter . . . creates a complex situation. I have decided several years ago not to have anything to do with initiatives officially involving the Italian government, whether in Italy or through the Embassy or consulates of Italy in the US. After 1945, the Italian government has never expressed regret or apologized to those like myself who because of racial laws were deprived of academic positions. . . . In recent years, the Italian government has evidenced an inclination to exploit Italian émigrés like myself for what I consider to be propaganda purposes. I have no wish to lend myself to such public relations operations for any government, including my own.[102]

Despite his desire not to be exploited by his government, Luria relished the role of public spokesman for science, and regularly testified before Congress about science funding and participated in forums on the place of science in American life.[103]

In his later years, Luria moved into a less rigorous mode of scientific activity. Although he continued to publish laboratory results with his fellow researchers, his rate of publication slowed to approximately one article a year between 1977 and 1983, the last year he published in a scientific journal.[104] Michael Weiss, his last graduate student, graduated from MIT in 1980.[105] From 1979 to 1985, Luria served on the institute's Equal Opportunity Committee, working to recruit more women and minorities as graduate students and faculty.[106] In 1984, Luria agreed to serve as a senior scientist for the Repligen biotechnology company in Needham, Massachusetts.[107] He maintained this position even after he officially retired from MIT in 1985.

Luria's last public pronouncement on a controversial scientific issue came soon after his retirement. In the debate over the proposed Human Genome Project (HGP), Luria argued that the scientific arguments in favor of the HGP did not outweigh the potential for social and political abuse.

By virtue of his eminence in the scientific community and his personal relationship with many of the key players, Luria felt justified in criticizing the proposed project.

One of the earliest public proposals for sequencing the entire human genome came from Luria's student and colleague Renato Dulbecco.[108] In March 1986, Dulbecco wrote an opinion column for *Science* linking cancer research to genomic research. He claimed that the next major useful step in the conquest of cancer would be to sequence a cellular genome, and the human genome would yield the most practical knowledge. With it, scientists could examine the differences between normal and cancerous cells and determine the cause of—and ultimately a possible treatment for—cancer.

In addition, Dulbecco argued, "Knowledge of the genome and availability of probes for any gene would be . . . crucial for progress in human physiology and pathology outside cancer." He proposed a national or even international initiative that would be on the scale of the space program. He acknowledged that there were many practical and technical problems involved with such a huge effort, but none was insurmountable. This type of project was necessary not only to close "one of our most challenging chapters in biomedical research" but also for the future of humanity. Hyperbolically, he wrote, "The sequence of the human DNA is the reality of our species, and everything that happens in the world depends on those sequences."[109]

Scientists and politicians had been discussing the possibility of a government-funded Human Genome Project for approximately a year when Dulbecco's article appeared. The United States Department of Energy (DOE) and the National Institutes of Health (NIH), as well as the National Research Council and the Office of Technology Assessment all prepared budgets, evaluated proposals, and discussed the feasibility of a Big Science type project for biology. Walter Gilbert, Robert Sinsheimer, and other prominent biologists in Luria's orbit immediately signed on to support the project in Senate budget hearings. Others, such as David Baltimore, expressed serious reservations over whether the project was technically possible and whether it would take resources away from other biological researchers. However, Baltimore was soon convinced that since computing technology was advanced enough, and the national research budget large enough, the Human Genome Project would be a boon for American biology and science in general.

Late in 1988, the NIH established an Office of Human Genome Research with Luria's student James D. Watson at its head. Although Watson was already the director of the Cold Spring Harbor Laboratory, he took on this second job because he realized that "only once would I have the opportunity to let my scientific life encompass the path from double helix to the three million steps of the human genome."[110]

The Human Genome Project had thus been a topic of public discussion for several years when Daniel Koshland Jr., the editor of *Science*, published an editorial on the "Sequences and Consequences of the Human Genome" in October 1989. Koshland outlined the "clear" medical benefits of sequencing the human genome, suggesting that causes and cures for many types of mental illness, Alzheimer's disease, cancers, and heart disease would soon follow. These medical cures would also heal social ills, since the problem of the mentally ill homeless was an acute one in the late 1980s. He dismissed any fears of coercion or loss of privacy with the assertion, "No individual should be forced to obtain genetic information but none should be denied information either."

Koshland also dismissed the fears of those who felt the project would justify discrimination and eugenic sterilization policies, by proposing all of which could be controlled through legislation. Although "we must be vigilant about ethical concerns," he cautioned vocal critics of the project not to be "paralyzed by outlandish scenarios." He ended the editorial with a forceful call to proceed with the sequencing of the human genome. While "there are immoralities of commission that we must avoid," the human genome represented "a great new technology to aid the poor, the infirm, and the underprivileged." Couched in these terms, the Human Genome Project could be the ultimate social welfare program, with obvious safeguards in place and only the greater good of mankind as its goal.[111]

Luria was disturbed by Koshland's editorial, as well as by much of the language used to describe the proposed Human Genome Project. He wrote a letter to *Science*, which was published the next month.[112] All of Luria's public roles converged in this short letter: eminent biologist, public statesman of science, social activist, political critic, defender of democracy, and beneath it all, Jewish refugee from Nazi Europe. Luria worried that the HGP had "been promoted without public discussion by a small coterie of power-seeking enthusiasts" who made promises of disease cures in order to fund basic research.[113] He charged his "friend, the usually level-headed editor of *Science*" with "befuddled" thinking, and expressed fear that "the phantom promise of early diagnosis of a few hereditary diseases is being replaced in Koshland's editorial by hints of a eugenic program."[114] Luria did not object to the HGP on funding or technical grounds; indeed, his experience in the arena of cancer research may have made Dulbecco's initial argument a convincing one for Luria. Nevertheless, he felt that insufficient attention had been paid to the possibility of eugenic applications of the project, and felt that Koshland's rhetoric was a hint of dangerous territory ahead.

Luria felt that there was a "real danger" of an institution that would "invade the rights and privacy of individuals . . . under cover of beneficent eugenic intervention." More than twenty-five years after he first wrote about the potential social and political abuses of "genetic surgery," Luria was frightened by what he saw as an impetuous rush into a large-scale scientific and technological program. He concluded his letter with an impassioned question. "Will the Nazi program to eradicate Jewish or otherwise 'inferior' genes by mass murder be transformed here into a kinder, gentler program to 'perfect' human individuals by 'correcting' their genomes in conformity, perhaps, to an ideal, 'white, Judeo-Christian, economically successful' genotype?"[115] Luria was not the only one expressing fears of a new eugenics from the Human Genome Project, but his particular life history and role in the scientific community made his hesitation significant.[116]

Koshland's published response was respectful and calm. He tried to reassure his "friend" that he agreed that the proposed Human Genome Project "requires special vigilance in the area of ethics," but cautioned him to remember, "unreal scenarios advance neither the cause of ethics nor that of science." Koshland remained convinced that the knowledge obtained by the project would lead to drugs that would help a wide range of illnesses, and that legislation before the fact would protect against any possible abuses by scientists or politicians.[117]

Luria's exchange with Koshland generated a private response from David Baltimore, who took issue with Luria's emotional stance and cautioned him to "not throw away the baby with the bath." Baltimore was confident that the Human Genome Project would "bring great benefits in terms of science" and that the role of scientists should be "to speak out clearly saying that the Genome Project is science and that science always has potential for good or evil." Despite Luria's "misreading" of the editorial, Baltimore felt that the Genome Project was good science and "should be discussed somewhat less emotionally than in your letter."[118] Baltimore had originally opposed the Human Genome Project on technical and financial grounds, but by 1989 he was a firm supporter of the project, especially with Watson's commitment to paying close attention to the ethical and social aspects of the sequence in place.[119] Watson interpreted Luria's opposition to the HGP as further evidence that Luria "cared more about politics than science."[120] Watson and Luria had not had a close personal relationship since the mid-1950s, and this disagreement only served to increase the distance between them.

Watson may have misunderstood the nature of Luria's attitude toward science and politics. Luria was convinced that there was a practical way for science and scientists to serve a modern democratic society, and he was passionately committed to finding ways that the rationality of science could ensure that all citizens in American society had equal opportunities. His conceptions of politics and science were born out of his experiences in each arena, and he believed that both provided the tools for achieving a society that would be free of inequality, oppression, and needless suffering. If his formulations seem naïve or unsophisticated, it may be because his perspective was as an activist rather than a theorist. In the aftermath of Vietnam and Watergate, Luria recognized that both the American democratic system and the role of science in American society were still in flux, but he was convinced that the way to a better world lay in the intersection of the two. Despite his fear of the abuse of biological ideas, Luria had faith that the power of scientific rationality to safely guide a democratic society to its full potential would ultimately prevail.

In the late 1980s, Luria was diagnosed with prostate cancer, which slowly spread throughout his body, causing him considerable pain. As they did in that eventful week in October 1969, Luria's public scientific and political roles converged in the months before his death on February 6, 1991. The threads of bacteriophage research, political activism, and government surveillance persisted until the very end of his life.

In December 1990, molecular biologist John Cairns wrote to Luria, asking for his response to Cairns's recent experimental results that seemed to contradict Luria's 1943 conclusions about spontaneous mutations from the fluctuation test. Luria died before they could meet in person to review the implications of Cairns's research, but it is a testament both to Luria's stature in the scientific community and the importance of his contributions that Cairns sought him out, nearly fifty years after Luria was inspired to design the fluctuation test by watching a colleague play the slot machine in Bloomington, Indiana.[1]

Even in his weakened physical condition, Luria was still committed to mobilizing his fellow academics for peace. In the fall of 1990, President George H. W. Bush sent American troops to the Persian Gulf in preparation for an international military response to Saddam Hussein's August invasion of Kuwait. In December, Luria and other members of the Temporary Faculty Committee for Peace in the Persian Gulf asked to meet with Massachusetts governor William Weld. They explained to the newly elected governor that as educators, "We want to bring Massachusetts' youngsters back alive—to teach them and train them to be useful citizens—not dead in body bags."[2]

On January 24, 1991, the FBI responded to a security request from the White House. The Secret Service was doing a routine background check on all of the twenty nominees for the National Medal of Science. Luria was to be honored at a ceremony in September "for a lifetime devoted to applying genetics to viruses and bacteria, and for guiding the development of generations of students who have helped create the modern power of molecular biology."[3] The FBI's two-page report indicated that Luria "was the subject of security investigations by the FBI during the 1940's and 1950's, 1960, 1967 and 1968." It also described a number of Luria's anti-Vietnam protests from the early 1970s. "No arrest record was located," and Luria was granted security clearance to attend the awards ceremony.[4]

Luria spent most of January 1991 in and out of the hospital. He suffered a heart attack and died at home on February 6.[5] The next day, newspapers all over the world carried obituaries for Salvador Edward Luria. Although the primary focus was on his Nobel Prize–winning research, they also mentioned his political activism, his blacklisting, and his long teaching career.[6]

Hundreds of condolence letters from colleagues, students, and friends arrived at the Lurias' home on Peacock Road in Lexington. His former students told Zella about how he was a powerful intellectual role model.[7] One student recalled how biology graduate students at MIT believed two things: if Luria told you something, it was true, and he was "one of the last people around to have an overview of all of biology."[8] Historian of science Gerald Holton praised him as "one of the last of the giants who changed the modern biological sciences. But in addition he was a major civilizing force on all those near him."[9] Renato Dulbecco told Zella that he wrote to her "with the same spirit I would do it if you were my mother. Because Salva was my father, who made me what I am now."[10] James Darnell and James Watson expressed similar filial feelings.[11]

Watson put his political disagreements with Luria aside when he paid public tribute to Luria in an obituary published in *Nature*. He described Luria's scientific talents, and his many contributions to modern biology through his bacteriophage research. He praised him as "a teacher of the first rank" and "an exceptionally talented writer." Watson mentioned that Luria was "a human of passionate political beliefs," and briefly mentioned their disagreement over "whether our societies could find the means to protect the victims of unjust throws of the genetic dice." He described how Luria "strove at all times to maintain high standards in both science and the human behavior that makes it possible," and concluded by noting "By the time he died, . . . there were few who did not feel better by being in his presence."[12]

Luria's body was cremated, and there was a private family service. Zella requested that donations in his memory be made to the American Friends Service Committee Peace Education Division, the Resource Center for Non-Violence, and the Children's Defense Fund. In a letter to Luria's friends, she expressed gratitude for reminders of Luria's "concern for fairness for students, scientists, and especially people with little power to fight for their own rights and needs in academia." She invited them to "celebrate Salva's life, his humanity, his struggle for understanding life and its biophysical basis, his sense of deep and personal fulfillment at having helped to build what he believed to be the best Biology Department in the country, his driving need to see justice done, his struggle for a peaceful, democratic world, his real interest in knowing people unlike himself, and his love of his family, friends, and co-workers."[13]

On April 1, 1991, friends and colleagues from all parts of Luria's life gathered to honor his memory at a public service at MIT. Boris Magasanik represented the MIT biology community, and Howard Zinn and Philip Morrison spoke for the academic peace movement. Although Luria did not observe Jewish customs, his close friend Bernard Levy recited Kaddish, the Jewish prayer for the dead.[14]

The scientific community honored Luria's memory in the months after his death as well. At the April 28 meeting of the Council of the National Academy of Sciences, Luria's death was noted with sadness. The members of the council passed a resolution praising Luria "for his wit and humanism as well as his scientific accomplishments." The short obituary essay focused mainly on the details of Luria's early years in the United States but also devoted considerable attention to his political activities. It concluded, "If Salvador Luria's strong social conscience made him controversial, his delightful sense of humor, trustworthiness, and dedication to teaching earned him nothing but love and respect."[15] In June, David Baltimore announced that the Rockefeller University was awarding Luria a posthumous honorary degree.[16] Although Luria had asked her to continue to live her own life instead of playing the role of his widow, Zella traveled to Washington, D.C., to accept Luria's National Medal of Science in September.[17] In the spring of 1992, MIT created an endowed professorship in Luria's memory and named Phillip Sharp, Luria's successor as the director of the Center for Cancer Research, as the first Salvador E. Luria Professor.[18] The biology department established prizes and lectures in Luria's memory, and more recently, the main meeting space at MIT's Koch Institute, the successor to the Center for Cancer Research, was named the Salvador E. Luria Auditorium.[19]

From the vantage point of late 2021, after nearly two years in which we confronted the devastating effects of a global pandemic coupled with widespread American rejection of scientific expertise as well as a painful reckoning with systemic racism, revisiting Luria's life shows us just how heartbreakingly far we have come from a time when immigrants were seen as valuable contributors to American society, scientific knowledge was respected, and scientific achievements were sources of national pride and optimism. But Luria's life story should also give us hope. This story reminds us that science, especially research that has the possibility of affecting human health, always

Salvador Edward Luria, 1912–1991. Photo courtesy of Richard Howard.

takes place within a constellation of political, social, and cultural pressures and concerns. This story reminds us that although state control of science can be dangerous, scientists have embraced the chance to help solve social problems. This story reminds us that Americans of all types have continuously fought for civil rights and social justice within their communities. This story reminds us that individual scientists have always had the courage to stand up for what they believe is right, both with regard to the uses of their research and with regard to what they want to change in American society. Luria was a brave and passionate spokesman for science who advocated for peace, equality, and opportunity at a particular time in American history, but he was not an anomaly. There are other Lurias out there. I hope this book inspires them to follow his lead.

ACKNOWLEDGMENTS

I have been thinking about Salvador Luria since 1997, when I was a first-year graduate student in Everett Mendelsohn's seminar on the history of biology. I am thankful to him, Allan Brandt, and the late Sam Schweber for guiding me through the dissertation process, and to my undergraduate advisors Robert Kohler, M. Susan Lindee, and Charles Rosenberg for preparing me to think historically and critically about American science.

Other scholars have been generous with their advice, time, and resources over the years, particularly Nathaniel Comfort, Angela N. H. Creager, Lara Freidenfelds, Michael Gordin, and Audra Wolfe. Special thanks go to David Kaiser for making the connection with Katie Helke. I found a supportive community of scholars at UCLA, and I am grateful to Jean-François Blanchette, Anne Gilliland, Russell Johnson, Gregory Leazer, and Marcia Meldrum, as well as to Wen Hsu and the legendary 2020 MLIS cohort, for their encouragement as I became an archivist while in the final stretch of this project.

I was fortunate to meet Zella Hurwitz Luria as I was researching Luria's life, and I enjoyed every moment I spent with her. She was a kind, warm, and insightful person and I am sorry she did not live to see this book published. Dan Luria has been equally gracious and encouraging, and I appreciate his sharing some of his favorite family photos with me. I am grateful to them and to everyone who took the time to share their memories of Luria with me over the years.

I acknowledge and thank the many archivists and librarians whose labor made the materials I used over the past twenty-plus years discoverable and accessible, especially Scott DeHaven, Robert Cox, Valerie Lutz, and Earle Spamer at the American Philosophical Society Library in Philadelphia, Dina Kellams, Bradley Cook, and the staff at the Indiana University Archive, Jeff Karr at the Center for the History of Microbiology/ASM Archives, Nora Murphy and the staff at the MIT Archive (now MIT Distinctive Collections),

Robert Chapel, Ellen Swain, and the eagle-eyed Katherine Nichols at the University of Illinois Urbana-Champaign Archive, Clare Bunce, Ludmilla Pollock, and Stephanie Satalino at the Cold Spring Harbor Laboratory Archive, and Dr. Erwin Levold at the Rockefeller Archive Center.

I am extremely thankful to Katie Helke, who felt that the time was right for The MIT Press to publish a book about Luria. It has been a pleasure working with her, Laura Keeler, and the rest of the staff at the Press. I appreciate the feedback from the three anonymous reviewers who read an early draft.

My friends have been wonderful throughout this project. Shira Berman and Julie Gruenbaum Fax were keen readers and editors, while Jeremy Horowitz and Arnon Z. Shorr gave me timely technical assistance. Taffy Brodesser-Akner and Sharon and Nick Merkin helped me navigate the publication process, and provided much needed distraction and encouragement whenever I needed them. The rest of my Los Angeles crew, including the Blumofes, Faxes, Gershovs, Hoffmans, Horowitzes, Kanefsky-Abrams, Katz-Bicks, Plutchok-Gordons, Rothmans, Smith-Finks, and Zelkhas, kept me sane and well fed over the years. For long-distance support, I can always count on Debra Bieler Klein, Shari Bursztyn, Aliza Fink and Michael Zatman, Cati and Daniel Freedman, Elizabeth Green, Rachel Goldberg, Tamar Kaplan Appel, Racheli Kraut, Sefi Kraut, Lisa Lederman, Dafna Michaelson Jenet, Rebecca Perlin and Danny Sadinoff, Jonathan Schloss, and David Wolf.

There is no way for me to adequately express my gratitude to my family. The Cohen-Lederman-Edelheit gang is generous with their love. I am so proud of my brothers, Micha and Isaac, who inspire me in countless ways, and I cherish the time I am able to spend with them along with Natalie, Annaelle, Assaf, Ayelet, and Amiad. My parents have supported me through thick and thin and through every single iteration of the Luria project. My mother, Barbara, read early drafts of this book, and introduced me to the satisfaction and adventure of a reading life. My father, Roger, started me on the path to being a historian when he told me stories from his mouth when I was a little girl and has wholeheartedly encouraged all of my intellectual pursuits. My daughters, Rebecca and Leah, are my favorite reading and baking and laughing companions and my greatest pride and joy. Like everything else, this book is for them.

NOTES

OCTOBER 1969

1. Weather forecast in *Boston Globe*, October 14–16, 1969.

2. Transcript of press conference in S. E. Luria "News" file, Massachusetts Institute of Technology Archives, MIT Museum, Cambridge, MA.

3. Lee, "3 Americans Get Nobel in Medicine"; McElheny and Black, "MIT Nobel Winner to Use Cash to End War."

4. Lyons, "Second H.E.W. Blacklist Includes Nobel Laureate."

5. Schrecker, *Many Are the Crimes*, 406.

6. Luria, "Modern Biology: A Terrifying Power."

7. Luria, "Paris. Lisbon. New York," manuscript [1981], Series III, Salvador E. Luria Papers, American Philosophical Society Library, Philadelphia. Hereafter SEL Papers, APS. Also available on the National Library of Medicine Profiles in Science page, accessed February 2022, https://profiles.nlm.nih.gov/spotlight/ql/catalog/nlm:nlmuid -101584611X144-doc.

8. Moore, *Disrupting Science*, 2.

9. Moore, *Disrupting Science*, 7.

10. Note from Daniel Luria on chap. 11 ("Emotions"), February 16, 1983, SEL Papers Series IIa, *A Slot Machine, A Broken Test Tube* Folder, SEL Papers, APS.

11. Luria, *Slot Machine*, 213, 215–216; Giuseppe Bertani, "Salvador Edward Luria (1912–1991)."

12. Moore, *Disrupting Science*, 7.

EARLY LIFE IN EUROPE

1. Author interview with Zella Hurwitz Luria, April 30, 2001; Luria, *Slot Machine*, 153.

2. Luria, *Slot Machine*, 209, 211.

3. Luria, *Slot Machine*, 14.

4. Roth, *History of the Jews of Italy*, 136, 490–508. See also Grilli, "Role of the Jews in Modern Italy," 60–81, 172–197, 260–280, on the political and economic roles Jews played in the emergence of the modern Italian state; and Hughes, *Prisoners of Hope*, on the effect of emancipation on Jewish identity.

5. Letter from Salvador E. Luria (hereafter SEL) to Alan Lourie, November 26, 1984, Lourie Folder, Series I, SEL Papers, American Philosophical Society Library (hereafter APS).

6. Epstein, "Mishpachat Luria (Luria Family)," 311.

7. Chaim Weizmann, *Trial and Error*, quoted in Michaelis, *Mussolini and the Jews*, 3; Roth, *History of the Jews in Italy*, 475.

8. H. Stuart Hughes cites Guido Bedarida's research on Jews in prominent positions in the first quarter of the twentieth century. Bedarida found that there were twenty-four Jewish members of the royally nominated senate, and 8 percent of university professors were Jewish. Bedarida, *Ebrei d'Italia*, cited in Hughes, *Prisoners of Hope*, 19–20.

9. Zuccotti, *Italians and the Holocaust*, 17–18.

10. On the sporadic persecution, see Levi, *Periodic Table*, 5; and Zuccotti, *Italians and the Holocaust*, 19. For examples of conversion attempts, see Levi-Montalcini, *In Praise of Imperfection*, 20; Segre, *Memoirs of a Fortunate Jew*, 30–45.

11. Levi, "Argon" in *Periodic Table*, 3–20.

12. Levi, "Argon" in *Periodic Table*, 5, 4.

13. Levi, "Jews of Turin." Reprinted as Appendix I in Anissimov, *Primo Levi: Tragedy of an Optimist*, 409.

14. Levi, "Jews of Turin," 409–412.

15. Author interview with Zella Hurwitz Luria; Luria, *Slot Machine*, 11–12, 154, 198–199.

16. See, for example, a letter from September 1946, Giuseppe Luria Folder, Series I, SEL Papers, APS.

17. Luria, *Slot Machine*, 153.

18. Note from SEL to Miriam Schurin, July 12, 1960, Brandeis University Correspondence Folder, Series I, SEL Papers, APS.

19. Hughes, *Prisoners of Hope*, 162.

20. Luria, *Slot Machine*, 12. This is in contrast to the cohort a few years above him, who became the leaders of the intellectual resistance, as discussed below.

21. Anissimov, *Primo Levi*, 26.

22. He considered himself a "non-naturalist." Luria, "Early Days in the Molecular Biology," 23; Luria, *Slot Machine*, 13.

23. Blatt, "Battle of Turin"; Pugliese, *Carlo Rosselli*, introduction and chap. 4.

24. Blatt, "Battle of Turin," 43–44n142; Pugliese, *Carlo Rosselli*, 194.

25. Wilkinson, *Intellectual Resistance in Europe*, 204; Bobbio, *Trent'anni di storia della cultura a Torino*, 49.

26. Bobbio, *Trent'anni*, 48. Translation by RES.

27. Luria, *Slot Machine*, 13.

28. Wilkinson, *Intellectual Resistance*, 195.

29. Zuccotti, *Italians and the Holocaust*, 28.

30. Luria, *Slot Machine*,14.

31. Luria, *Slot Machine*, 14–15.

32. Amprino, "Giuseppe Levi," 17; Olivo, "Giuseppe Levi."

33. Ginzburg, *Family Sayings*, 36; Levi-Montalcini, *In Praise of Imperfection*, 48.

34. Ginzburg, *Family Sayings*, 19, 30.

35. Luria Curriculum Vitae, September 1940, Salvador Luria Folder, Rockefeller Archive Center Record Group 1.1, Series 200, Box 133, Folder 1641, Rockefeller Archive Center, Sleepy Hollow, NY. Hereafter SEL-RAC.

36. Levi-Montalcini, *In Praise of Imperfection*, 65–66.

37. Letter about Salvatore Luria from Giuseppe Levi, March 17, 1938, p. 2, Università degli studi di Torino Folder, Series IIb, SEL Papers, APS; Luria, *Slot Machine*, 16.

38. Ginzburg, *Family Sayings*, 86–87; Hughes, *Prisoners of Hope*, 95–96; Zuccotti, *Italians and the Holocaust*, 28–29; Blatt, "Battle of Turin," 31–32.

39. Ginzburg, *Family Sayings*, 86–77; Levi-Montalcini, *In Praise of Imperfection*, 77.

40. Luria, *Slot Machine*, 172; quote from Levi-Montalcini, *In Praise of Imperfection*, 77.

41. Levi-Montalcini, *In Praise of Imperfection*, chap. 6; Dulbecco, *Scienza, vita e avventura*, 46–48; Kevles, "Renato Dulbecco and the New Animal Virology," 411–412.

42. Levi Montalcini, *In Praise of Imperfection*, 112–113, 137–138.

43. Luria Curriculum Vitae, September 1940, SEL-RAC.

44. "Servizi, Promozioni e Variazioni," Italian Army documentation provided in U.S. visa application, Salvador E. Luria Immigration and Naturalization Service File. Hereafter SEL-INS. Luria, *Slot Machine*, 17–18; author interview with Zella Hurwitz Luria, April 30, 2001.

45. Kay, *Molecular Vision of Life*, 111.

46. Luria, *Slot Machine*, 62.

47. Segré, "Enrico Fermi," in *Dictionary of Scientific Biography*, 578–579; Segré, *Enrico Fermi: Physicist*, 48–53; Cronin, *Fermi Remembered*.

48. Luria, *Slot Machine*, 19.

49. Segré, *Enrico Fermi*, 23.

50. Kay, *Molecular Vision of Life*, 133.

51. Erwin Schrödinger took up this tantalizing but ultimately incorrect model in his famous book *What Is Life?* See Olby, *Path to the Double Helix*; Judson, *Eighth Day of Creation*; Kay, "Quanta of Life"; Yoxen, "Where Does Schrödinger's '*What Is Life?*' Belong."

52. For general biographical information on Delbrück, see Stent, "Max Delbrück, 1906–1981" *Genetics* 101; Kay, "Conceptual Models and Analytical Tools"; and Fischer and Lipson, *Thinking About Science*.

53. Delbrück and Delbrück, "Bacterial Viruses and Sex," 49, cited in Judson, *Eighth Day of Creation*, 51.

54. Max Delbrück oral history, 1978, p. 57, Caltech Archives, cited by Kay, *Molecular Vision of Life*, 135.

55. Kay, "Conceptual Models and Analytical Tools," 213.

56. Muller, "Variation Due to Change," 48–49.

57. Summers, "How Bacteriophage Came to Be Used," 262; Summers, "Inventing Viruses," 32.

58. Luria, *Slot Machine*, 20.

59. See Fischer and Lipson, *Thinking About Science*, 112–114; Kay "Quanta of Life," 10; and Summers, "How Bacteriophage Came to Be Used," on how Delbrück stumbled into Emory Ellis's lab after not getting anywhere with *Drosophila*.

60. Luria, *Slot Machine*, 21.

61. van Helvoort, "History of Virus Research," 185–186; Fenner, "Poxviruses," 2–3; Hughes, *Virus: A History of the Concept*, especially chap. 6; Collard, *Development of Microbiology*; Waterson and Wilkinson, *Introduction to the History of Virology*, chaps. 7–9.

62. For a detailed account of d'Herelle's life work, see Summers, *Félix d'Herelle*, especially chap. 4, on the discovery of bacteriophage. The other side of the story is told in Twort, *In Focus, Out of Step*. A short documentary review and analysis of the debate is Duckworth, "Who Discovered Bacteriophage?"

63. Twort, "An Investigation of the Nature of Ultra-Microscopic Viruses." Reproduced in Twort, *In Focus, Out of Step*; quoted extensively in Duckworth, "Who Discovered," 794–795.

64. Duckworth, "Who Discovered," 794.

65. Twort, "An Investigation," 1243.

66. Félix d'Herelle, "On an Invisible Microbe Antagonistic to the Dysentery Bacillus," translated as Appendix in Summers, *Félix d'Herelle*, 185–186. It is unknown whether he had seen Twort's *Lancet* article at the time.

67. Summers, *Félix d'Herelle*, 59.

68. Summers, *Félix d'Herelle*, 64, 77.

69. van Helvoort, "Construction of Bacteriophage"; van Helvoort, "History of Virus Research," 188; Waterson and Wilkinson, *Introduction to the History*, 89–93; Sankaran, "When Viruses Were Not in Style," 191–192; Summers, "Inventing Viruses"; Summers, "Félix Hubert d'Herelle."

70. Lederman and Tolin, "OVATOOMB." See also van Helvoort, "History of Virus Research," 202. For a discussion of the history of *Drosophila* research, see Kohler, *Lords of the Fly*.

71. Waterson and Wilkinson, *Introduction to the History*, chap. 7; van Helvoort "History of Virus Research," 186.

72. Waterson and Wilkinson, *Introduction to the History*, 17–18.

73. Lederman and Tolin, "OVATOOMB," 253. For an analysis of the history of Wendell Stanley's work on the tobacco mosaic virus, see Kay, "W. M. Stanley's Crystallization of the Tobacco Mosaic Virus," and Creager, *Life of a Virus*.

74. Zimmer, "Target Theory," in Cairns, Stent, and Watson, *Phage and the Origins of Molecular Biology*, 33–42, at 36. Hereafter *PATOOMB*.

75. Waterson and Wilkinson, *Introduction to the History*, 98.

76. Luria, *Slot Machine*, 64–69.

77. Luria, "Early Days in the Molecular Biology," 22.

78. Michaelis, *Mussolini and the Jews*, 152–153.

79. Zuccotti, *Italians and the Holocaust*, 36.

80. Giuseppe Levi and Rita Levi-Montalcini were in the same situation. Both chose to remain in Italy and spent most of the war in hiding. Levi-Montalcini, *In Praise of Imperfection*, Section II: "Difficult Years"; Ginzburg, *Family Sayings*, 120.

81. Luria's parents and brother survived World War II by hiding with sympathetic friends. Luria, *Slot Machine*, 22.

82. The Italian letter, dated March 25, 1938, is from Luria's application to go to Copenhagen to study at the Institute of Theoretical Physics at the university there. The English letter is a more general one. Letters of Introduction from Enrico Fermi, February 14 and March 25, 1938, Enrico Fermi Folder, Series I, SEL Papers, APS.

83. Luria, *Slot Machine*, 69–70.

84. Luria, *Slot Machine*, 70.

85. Wollman, Holweck, and Luria, "Effect of Radiations on Bacteriophage C16."

86. Luria, *Slot Machine*, 23–24.

87. Luria, *Slot Machine*, 173. For a discussion of the organization of the Italian anti-Fascists in Italy in the 1930s, see Blatt, "Battle of Turin," and Pugliese, *Carlo Rosselli*.

88. Luria, *Slot Machine*, 174–175.

89. Fosdick, *Story of the Rockefeller Foundation*, 276–277. See also Abir-Am, "Rockefeller Foundation and Refugee Biologists."

90. Handwritten note from Holweck to Henry Miller, March 5, 1940, SEL-RAC.

91. Letter from Miller to Holweck, April 18, 1940, SEL-RAC.

92. Application for Immigration Visa Number 1895, American Consulate at Paris, France, June 5, 1940, SEL-INS.

93. Luria, *Slot Machine*, 24. On the fate of French Jews, see Lachmanovich, "Holocaust of the French Jews."

94. Luria, *Slot Machine*, 27.

95. Luria, "Mutations of Bacteria and of Bacteriophage," in Cairns, Stent, and Watson, *PATOOMB*, 173–179, at 177.

96. Certification from the Comissao Portuguesa de Assistencia aos Judeus Refugiados, August 27, 1940, Comissao Portuguesa de Assistencia aos Judeus Refugiados Folder, Series IIb, SEL Papers, APS.

97. Luria, *Slot Machine*, 29; Immigration Visa 5258, admitted September 12, 1940, SEL-INS.

BECOMING AN AMERICAN BIOLOGIST

1. S. E. Luria, "Domestic Cooperation," n. d., Series III, SEL Papers, APS, p. 2.

2. Letters of recommendation from Max Delbrück, Milislav Demerec, Enrico Fermi, Michael Heidelberger, Frank Meleny, Hermann Muller, John Northrop, Charles Packard,

L. J. Stadler, and W. H. Woglom to the John Simon Guggenheim Foundation, October 1941, Salvador E. Luria File, Archive of the John Simon Guggenheim Memorial Foundation. Hereafter Guggenheim Foundation Archives.

3. Letter from Frank Exner to SEL, October 15, 1969, Nobel Prize Folder #1, October 15–16, A-M, SEL Papers, APS.

4. "Alien Registration Form," November 20, 1940; "Declaration of Intention," May 7, 1941, SEL-INS.

5. Luria, *Slot Machine*, 32.

6. Gormley, "Scientific Discrimination," 36; "Biographical/Historical Information," Emergency Committee in Aid of Displaced/Refugee Scholars Records Manuscripts and Archives Division, New York Public Library, http://archives.nypl.org/mss/922#overview.

7. Luria, *Slot Machine*, 31; Dazian Foundation for Medical Research and American Cancer Society Folders, SEL Papers, APS; Salvador Luria File, Rockefeller Foundation Papers, Rockefeller Archive Center.

8. Luria, *Slot Machine*, 70; Luria and Exner, "Inactivation of Bacteriophages"; Exner and Luria, "Size of Streptococcus Bacteriophages"; Luria and Exner, "On the Enzymatic Hypothesis of Radiosensitivity."

9. Abir-Am, "Rockefeller Foundation and Refugee Biologists," 230; Fischer and Lipson, *Thinking About Science*, 127.

10. Luria, *Slot Machine*, 33; Waterson and Wilkinson, *History of Virology*, 104.

11. Stent, "That Was the Molecular Biology That Was," 393. Reproduced in Cairns, Stent, and Watson, *PATOOMB*.

12. Kay, "Conceptual Models and Analytical Tools," 237–238. Luria was known as "Lu" among most biologists and phage researchers of the time—his correspondence from that period is most often addressed to Lu and he signed most letters to colleagues that way.

13. Fischer and Lipson, *Thinking About Science*, 130.

14. Berry, Inokuti, and Rau, "Ugo Fano, 1912–2001," 7.

15. The history of the Cold Spring Harbor Laboratory (CSHL) is described in Comfort and Glass, *Building Arcadia* (unpublished manuscript courtesy of Nathaniel Comfort), as well as in Comfort, *Tangled Field*. Several accounts of life at CSHL during this time period can be found in John Inglis, Sambrook, and Witkowski, *Inspiring Science*.

16. Mullins, "Development of a Scientific Specialty," 72.

17. Kay, "Conceptual Models and Analytical Tools," 213.

18. Wollman, Holweck, and Luria, "Effect of Radiation," 936. Earlier historical narratives about bacteriophage research argued that Luria and Delbrück were not

interested in biochemistry, despite their decision to publish in this journal and pursue biochemical explanations of inactivation. See especially Watson, *Double Helix*, 17–18. Luria addressed this issue in his recollections "Early Days in the Molecular Biology of Bacteriophage." The disciplinary rivalry between biochemistry and molecular biology continues to this day. See Witkowski, *Inside Story*, for several primary accounts of the "unacknowledged" contributions of biochemists, and the special issue of the *Journal of the History of Biology* (volume 29, September 1996) on "The Tools of the Discipline: Biochemists and Molecular Biologists," edited by Soraya de Chadarevian and Jean-Paul Gaudillière.

19. Delbrück and Luria, "Interference Between Bacterial Viruses: I. Interference Between Two Bacterial Viruses Acting Upon the Same Host," 111; Luria and Delbrück, "Interference Between Inactivated Bacterial Virus and Active Virus."

20. The nomenclature for identifying phage strains is a bit confusing, since there was no consistency until the "Phage Treaty" of 1944 discussed later, in which the community of phage workers agreed to use the same labels, T1 through T7, to refer to the *E. coli* phages. Until 1944, Luria, Delbrück, and others used the Greek symbols that they had on their typewriters to refer to the different strains. See Luria, *Slot Machine*, 74; Kay, *Molecular Vision of Life*, 245; and Brock, *Emergence of Bacterial Genetics*, 127.

21. Delbrück and Luria, "Interference Between Bacterial Viruses," 136–137.

22. George Beadle and Edward Tatum published their results with *Neurospora*, the famous "one gene-one enzyme" theory in 1941.

23. Kay, *Molecular Vision*, 110–112.

24. Delbrück and Luria, "Interference Between Bacterial Viruses," 137–138.

25. Luria, "Plan for Work," Salvador E. Luria File, Guggenheim Foundation Archive, 1–2.

26. Luria, "A Reminiscence on the Department of Genetics, Carnegie Institute of Washington, Cold Spring Harbor," n.d., Series III, SEL Papers, APS.

27. Confidential Report on Candidate for Fellowship on Luria by Demerec, October 20, 1941, Salvador E. Luria File, Guggenheim Foundation Archive.

28. Confidential Report on Candidate for Fellowship on Luria by Muller, (date unclear) 1941, Salvador E. Luria File, Guggenheim Foundation Archive.

29. Memberships among those listed in Luria's entry in the 1949 *American Men of Science*, 1546.

30. On the "fragmentation of biology" in the early twentieth century, see Appel, "Organizing Biology."

31. Anderson, "Electron Microscopy of Phages," 64.

32. Anderson, "Electron Microscopy," 63–64; Rasmussen, *Picture Control*, 25–26, 48; Kruger and Gelderblom, "Helmut Ruska"; van Helvoort and Sankaran, "How Seeing Became Knowing," 130–132.

33. Rasmussen, *Picture Control*, 29, 46–48, 51.

34. Anderson, "Electron Microscopy," 64.

35. Rasmussen, *Picture Control*, 51, 57.

36. Anderson, "Electron Microscopy," 64; Luria, *Slot Machine*, 35.

37. Luria, *Slot Machine*, 20, 35–36, 67.

38. Luria and Anderson, "Identification and Characterization of Bacteriophages," 128–130.

39. Luria, "Plan for Work," p. 3, Salvador E. Luria File, Guggenheim Foundation Archive.

40. Rasmussen, *Picture Control*, 52.

41. Luria, Delbrück, and Anderson, "Electron Microscope Studies," 60.

42. Luria, Delbrück, and Anderson, "Electron Microscope Studies," 61–62.

43. Rasmussen, *Picture Control,* 57.

44. Luria, Delbrück, and Anderson, "Electron Microscope Studies," 65–66.

45. Anderson, "Electron Microscopy," 71.

46. Luria, Delbrück, and Anderson, "Electron Microscope Studies," 64. This article and this particular paragraph, are also discussed briefly in Kay, *Molecular Vision*, 245, although she mentions only Delbrück and Anderson with respect to this work, and by Creager in *Life of a Virus*, 205–208.

47. See the comment by Altenburg, at the end of Luria's 1951 paper from Cold Spring Harbor Symposia, saying, "I suggest that the conclusion to be drawn from Dr. Luria's experiments is that the reproduction cycle of viruses conforms with that of organisms in general, and that therefore viruses are to be considered organisms." S. E. Luria, "Frequency Distribution of Spontaneous Bacteriophage Mutants," 463–470.

48. A Free French Scientist, "Dr. F. Holweck," *Nature* 149, no. 3771 (February 7, 1942): 163.

49. Rosenblum and Luria, "Fernand Holweck, 1889–1941," 330.

50. Salvador Luria acceptance letter, September 24, 1942, in Salvador Luria Folder, Herman B. Wells Papers, University Archives, Indiana University, Bloomington. Hereafter Wells Papers.

51. Luria, "Mutations of Bacteria and of Bacteriophage," 174–175. See also Luria, *Slot Machine*, 75–79.

52. Luria and Delbrück, "Mutations of Bacteria from Virus Sensitivity," 502.

53. This description of the experiment is based on Luria and Delbrück, "Mutations of Bacteria" as well as the explanation in Luria, *Slot Machine*, 75–79, and Judson, *Eighth Day of Creation*, 55–56.

54. See Stent, *Molecular Genetics*, 157; and Brock, *Emergence of Bacterial Genetics*, 62–63.

55. Luria, *Slot Machine*, 78–79.

56. Luria, "Mutations of Bacteria and of Bacteriophage," 174–175. He quotes from letters exchanged with Delbrück at the time, but they are no longer in existence in either of their archival collections. See letters between Delbrück and Luria from December 13 and 17, 1968, in the Max Delbrück Papers.

57. Luria, *Slot Machine*, 79.

58. Luria and Delbrück, "Mutations of Bacteria." Note that Luria is listed as the first author, although the footnote lists Delbrück's contribution first. Abir-Am, "Entre mémoire collective et histoire," has a similar observation on page 49.

59. Sapp, *Beyond the Gene*, chap. 4.

60. Letter from Luria to Tracy Sonneborn, June 14, 1943. Salvador E. Luria Folder, Tracy M. Sonneborn Papers, Lilly Library, Indiana University, Bloomington. Hereafter Sonneborn Papers.

61. Sonneborn had observed this phenomenon in *Paramecium* earlier that year. Letter from Sonneborn to Luria, June 17, 1943, Salvador E. Luria Folder, Sonneborn Papers.

62. Letter from Luria to Sonneborn, July 7, 1943, Salvador E. Luria Folder, Sonneborn Papers.

63. Letter from Luria to Sonneborn, July 7, 1943, Salvador E. Luria Folder, Sonneborn Papers.

64. Letter from Luria to Sonneborn, July 16, 1943, Salvador E. Luria Folder, Sonneborn Papers.

65. Brock, *Emergence of Bacterial Genetics*, 61.

66. Luria and Delbrück, "Mutations of Bacteria," 492.

67. Luria and Delbrück, "Mutations of Bacteria," 493.

68. Luria and Delbrück, "Mutations of Bacteria," 494.

69. Luria and Delbrück, "Mutations of Bacteria," 494–495. Emphasis in original.

70. Luria and Delbrück, "Mutations of Bacteria," 494.

71. Luria and Delbrück, "Mutations of Bacteria," 509.

72. Luria and Delbrück, "Mutations of Bacteria," 494.

73. Luria, "Mutations of Bacterial Viruses," 97.

74. Luria and Delbrück, "Mutations of Bacteria," 494.

75. Web of Science citation analysis performed by the author, May 13, 2020.

76. Brock, *Emergence of Bacterial Genetics*, 62–69.

77. Hershey, "Mutation of Bacteriophage"; Demerec and Fano, "Bacteriophage-Resistant Mutants"; Demerec, "Production of Staphylococcus Strains"; Demerec, "Origin of Bacterial Resistance to Antibiotics."

78. Joshua Lederberg and Edward Tatum, "Novel Genotypes in Mixed Cultures of Biochemical Mutants of Bacteria," 112–114; Andre Lwoff, "Some Problems Connected with Spontaneous Biochemical Mutations in Bacteria," 139–155; Francis J. Ryan, "Back-Mutation and Adaptation of Nutritional Mutants," 215–227; Arthur Shapiro, "Kinetics of Growth and Mutation in Bacteria," 228–235; Edward Tatum, "Induced Biochemical Mutations in Bacteria," 278–284, all in *Cold Spring Harbor Symposia on Quantitative Biology* 11 (1946).

79. Creager, "Adaptation or Selection?"

80. Oakberg and Luria, "Mutations to Sulfonamide Resistance."

81. See Braun, "Bacterial Dissociation."

82. Braun, "Studies on Bacterial Variation," 262. This was the first article on bacteria to appear in this general biology journal. See Smocovitis, "Organizing Evolution," 254, on how this journal was becoming increasingly focused on genetics, as opposed to evolution, in the 1940s.

83. Luria, "Recent Advances in Bacterial Genetics," 1. In *A History of Molecular Biology* (55), Michel Morange claims that this experiment "linked Darwinism and molecular biology once and for all."

84. Luria, "Recent Advances," 2.

85. Luria, "Recent Advances," 19.

86. Luria, "Recent Advances," 29.

87. Luria, "Recent Advances," 30.

88. Creager, "Mapping Genes in Microorganisms," 11.

89. Smocovitis, *Unifying Biology*; Creager, "Adaptation or Selection?"; O'Malley, "Experimental Study of Bacterial Evolution," esp. 319–329.

90. Witkin, "Chances and Choices," 3.

91. Witkin, "Chances and Choices," 6.

92. Witkin, "Genetics of Resistance to Radiation."

93. Luria, *Slot Machine*, 82; author interview with Zella Hurwitz Luria, January 13, 2000; author interview with Jerard Hurwitz, July 2, 2001.

94. Fischer and Lipson, *Thinking About Science*, 149.

95. Stent, "1969 Nobel Prize for Physiology or Medicine," 481.

96. The list of Luria's published works does not include any joint publications with Hershey, although they did develop a close friendship. See Hershey Folder and Condolence Letters Folder in SEL Papers, APS.

97. Luria, *Slot Machine*, 42.

98. Watson, "Growing Up in the Phage Group"; Luria, *Slot Machine*, 97–98.

99. Historical accounts include *PATOOMB*; Olby, *Path to the Double Helix*; and Judson, *Eighth Day of Creation*. Mullins, "Development of a Scientific Specialty" is a social network analysis also based on *PATOOMB*.

100. I use the term "young woman" because they are referred to as "Miss." Apparently, many young women and scientific wives acted as laboratory technicians for these scientists. Millard Susman's recollection of the phage course mentions his wife and several other workers in the "phage kitchen." Susman, "Cold Spring Harbor Phage Course," 1102.

101. Abir-Am, "First American and French Commemorations," 342–345.

102. Sloan, "Molecularizing Chicago,"387.

103. Letter from Max Delbrück to Salvador E. Luria, March 28, 1949, Max Delbrück Folder, SEL Papers, APS.

104. A large number of the letters in Luria's correspondence from the 1940s and 1950s are requests for phage stock or for advice on how to grow and plate it.

105. Luria, *Slot Machine*, 74. In a paper published in January 1945, Luria referred to several strains of bacteria and their viruses following "the convention of naming the mutant bacterial strains by the Roman letter corresponding to the original strain, followed by the Greek letter corresponding to the virus. . . . Strains isolated in the presence of the same virus, but different in some properties . . . are distinguished by subindexes. . . . Mutants obtained from other mutants are named by adding the Greek letter corresponding to the virus in the presence of which the new mutant has been isolated." Luria, "Mutations of Bacterial Viruses Affecting Their Host Range," 85.

106. Kay, *Molecular Vision*, 245. See Kohler, *Lords of the Fly*, chap. 5, on the Drosophila Exchange Service.

107. Kay, *Molecular Vision*, 245–246.

108. Fischer and Lipson, *Thinking About Science*, 157.

109. Comfort and Goldstein, *Building Arcadia*, 4.

110. Luria, "Mutations of Bacteria and Bacteriophage," 178. Unfortunately, that correspondence also did not survive in the Luria or Delbrück papers.

111. Even at Cold Spring Harbor, the bulk of the research in the genetics department revolved around work on penicillin. See Demerec and Luria, "The Gene," 117–119.

112. Fischer and Lipson, *Thinking About Science*, 157.

113. Comfort and Goldstein, *Building Arcadia*, 5, 7.

114. Kay, "Conceptual Models and Analytical Tools," 241.

115. Demerec, "Annual Report of the Biological Laboratory," 1945, 20.

116. Rasmussen, *Picture Control*, 8. See also Rasmussen, "Midcentury Biophysics Bubble."

117. Demerec, "Annual Report of the Biological Laboratory," 1948, 16.

118. Comfort and Goldstein, *Building Arcadia*, 9.

119. Susman, "Cold Spring Harbor Phage Course," 1101, 1103.

120. Luria, *Slot Machine*, 41; Watson, "Growing Up in the Phage Group," 240; Judson, *Eighth Day of Creation*, 64.

121. Watson, "Growing Up in the Phage Group," 242.

122. Taped interview with L. S. McClung, May 9, 1968, American Society for Microbiology Archive, Baltimore, MD; Luria, *Slot Machine*, 42; Kevles, "Renato Dulbecco and the New Animal Virology," 415–417; Baltimore, "Renato Dulbecco: A Biographical Memoir."

123. Luria and Dulbecco, "Genetic Recombinations." This article includes a mathematical appendix by Dulbecco, entitled "On the Reliability of the Poisson Distribution as a Distribution of the Number of Phage Particles Infecting Individual Bacteria in a Population." Submitted June 1948, 93.

124. Letter to GSA Members from R. E. Cleland, January 17, 1947, 1947 GSA Committee on Aid to Geneticists Abroad Folder, Box 4, Genetics Society of America Archives, American Philosophical Society Library, Philadelphia. Hereafter GSA Papers.

125. Ralph E. Cleland, "Report for 1947 of Committee on Aid to Geneticists Abroad," December 1947, p. 2, 1948 GSA Committee on Aid to Geneticists Abroad Folder, Box 5, GSA Papers.

126. Krementsov, "'Second Front' in Soviet Genetics," 240.

127. The extensive literature on the Lysenko controversy includes: Joravsky, *Lysenko Affair*; Medvedev, *Rise and Fall of T. D. Lysenko*; Paul, "War on Two Fronts";

Krementsov, "'Second Front' in Soviet Genetics"; Harman, "C. D. Darlington"; Wolfe, "What Does It Mean to Go Public?"; Wolfe, *Freedom's Laboratory*, chap. 1. Contributions to the *Journal of the History of Biology* special issue on the Lysenko controversy and the Cold War (volume 45, Fall 2012) include: deJong-Lambert and Krementsov, "On Labels and Issues"; Wolfe, "Cold War Context"; Selya, "Defending Scientific Freedom and Democracy"; Gordin, "How Lysenkoism Became Pseudoscience."

128. Wolfe, *Freedom's Laboratory*, 21.

129. Wolfe, *Freedom's Laboratory*, 24–25.

130. Paul, "War on Two Fronts"; Krementsov, "'Second Front' in Soviet Genetics"; Harman, "British and American Reaction to Lysenko"; Wolfe, "What Does It Mean to Go Public?" Wolfe, *Freedom's Laboratory*, 24–25.

131. Krementsov, "'Second Front' in Soviet Genetics," 240–244.

132. Letter from Muller to Stern, February 11, 1946, quoted in Krementsov, "'Second Front' in Soviet Genetics," 242.

133. It is unclear how Luria was selected to write the letter on behalf of this group of geneticists.

134. April 6, 1946, Draft in Herman Muller Papers; letter from K. Stern and others to J. B. S. Haldane, dated April 17, 1946, in the Stern Papers at the American Philosophical Society Library, Philadelphia, quoted in Krementsov, "'Second Front' in Soviet Genetics," 243. See Paul, "War on Two Fronts," 22–30, for a discussion of Haldane's torn loyalties and his need to defend Lysenko's theories to the outside world while he criticized them within the more private world of the Communist Party.

135. Letter from Luria to Muller, April 23, 1946, Muller Papers.

136. Krementsov, "'Second Front' in Soviet Genetics," 243.

137. Paul, "War on Two Fronts," 21–23.

138. Rabinowitch, "Purge of Genetics in the Soviet Union," is an excellent primary source summary of the events of August 1948.

139. This section is based on Selya, "Defending Scientific Freedom and Democracy."

140. Genetics Society of America Minute book for 1949, GSA Papers; Irwin, "Proceedings of Meetings," 24.

141. Letter to GSA Members from the Committee of Nine, August 25, 1950, GSA Committee to Combat Anti-Genetic Propaganda Folder, Box 7, GSA Papers. Emphasis in original.

142. Letter to GSA Members from the Committee of Nine, August 25, 1950, GSA Committee on Public Education and Scientific Freedom 1950 Folder, GSA Papers.

143. Statement by S. E. Luria, GSA Committee on Public Education and Scientific Freedom 1950 Folder, GSA Papers.

144. "Certificate of Naturalization" No. 6502470, January 10, 1947, SEL-INS.

145. Letters from Luria to Delbrück, January 6 and 24, 1947, Luria Correspondence, Delbrück Papers, Caltech.

146. Letter from Luria to Wells, January 10, 1947, from Wells to Luria, January 17, 1947, Salvador Edward Luria Folder, Wells Papers.

147. Luria, *Slot Machine*, 179.

148. Luria, *Slot Machine*, 180.

149. Luria, *Slot Machine*, 181. Luria wryly recalls, "The Indiana University Teachers Union never had more than twelve members, although our self-assurance made us look as if we were at least fifteen."

150. Indiana University Teachers Union Local 856, A.F.T., A.F. of L. Bulletin 4, January 1950.

151. Teachers Union Bulletin, May 1950, 2–3.

152. There is a substantial literature on the Progressive Party and on Henry Wallace's place in American history. See MacDougall, *Gideon's Army*; Markowitz, *Rise and Fall of the People's Century*; Walton, *Henry Wallace, Harry Truman, and the Cold War*; Culver and Hyde, *American Dreamer*; Kleinman, *World of Hope*.

153. Luria, *Slot Machine*, 182.

154. There was some confusion about the spelling of his name and if Salvatore, Salvador, and S. E. Luria (and occasionally Lurie) were all the same person. Memorandum to SAC (Special Agent in Charge) New York, from Director, FBI, January 18, 1950, FBI file on Salvador Edward Luria, File Number 100–367221. Hereafter FBI-SEL.

155. "Memorandum" from C. E. Heinrich to A. H. Belmont of the INS, November 8, 1950. FBI-SEL.

156. "Index to Informants" in Report from Indianapolis on Salvatore Edward Luria, July 7, 1950, 10. FBI-SEL. Of course, the names and other information about these informants are blacked out.

157. Report from Indianapolis on Salvatore Edward Luria, July 7, 1950, 1. FBI-SEL. The synopsis actually includes a typographical error, saying Luria started working at Indiana University in January 1942, when in fact it was 1943.

158. "Physical Description," Report from Indianapolis on Salvatore Edward Luria, July 7, 1950, 3. The characterization fits commonly held stereotypes about both Italians and Jews. FBI-SEL. As far as I can tell, Luria's Jewishness was not mentioned by informants

or FBI agents, but his foreignness was. Although he spoke English fluently and wrote it with great clarity, Luria's accent was in fact quite heavy, especially in the 1940s.

159. "Communist Party Activity" in Report from Indianapolis on Luria, July 7, 1950, 3–4. FBI-SEL. There is no evidence of such a letter about the American Federation of Teachers in the Presidential papers in the Indiana University Archives.

160. "Communist Party Activity" in Report from Indianapolis on Luria, July 7, 1950, 4, 6. FBI-SEL.

161. "Communist Party Activity" in Report from Indianapolis on Luria, July 7, 1950, 5. FBI-SEL.

162. "Communist Party Activity" in Report from Indianapolis on Luria, July 7, 1950, 6. FBI-SEL.

163. Lindgren, "Letter to the Editor," 2.

164. Genetics Society of America Minute book for 1949, GSA Papers.

165. "Leads," in Report from Indianapolis on Luria, July 7, 1950, p. 9. FBI-SEL.

166. Theoharis, "Wiretaps, Mail Openings and Break-Ins" in *Spying on Americans*, 94–132.

167. "Memorandum" from SAC, Indianapolis, to Director, FBI, October 2, 1950, and "Memorandum" from SAC, Springfield to Director, FBI, November 15, 1950. FBI-SEL.

168. "Memorandum" from SAC, Indianapolis, to Director, FBI, October 2, 1950; and "Administrative Page: Miscellaneous, Associates, Contacts" in Report from Springfield, November 7, 1950, 11. FBI-SEL.

169. During this period Luria did not publish an article in the magazine nor did they print any letters he may have written to the editor. However, he was mentioned in several articles about genetics and bacteria, and he may have served as a referee for these articles. His archived correspondence contains no letters to or from the editors of *Scientific American*.

170. "Letter" from SAC, NY to Director, FBI, October 31, 1950, 1–2. FBI-SEL. Despite the FBI's vigilance about protecting informants and the privacy of individuals named in other people's files, the censor did not black out Chambers's name.

171. Report from New York, March 14, 1952, 3. FBI-SEL. See Schweber, *In the Shadow of the Bomb*, 160–164, for a discussion of Bethe's opposition to the development and use of the hydrogen bomb.

172. Clippings from *Indiana Daily Student*, June 14, 1947, June 21, 1947, March 24, 1948, and October 16, 1948, Luria Clippings File, Indiana University Archives.

173. January 9, 1948, letter in Salvador Edward Luria Folder, Wells Papers; "News and Notes," *Science* 107 (March 12, 1948): 264–268, 264. On the circumstances of Slotin's death see Wellerstein, "Demon Core."

174. Press release from Columbia University Public Information Office, April 5, 1950, Columbia University Archives, New York. On the Jesup Lectures, see Cain, "Co-opting Colleagues," and Cain, "Columbia Biological Series."

175. L. C. Dunn to Nicholas Butler, May 26, 1937, cited in Cain, "Co-opting Colleagues," 214.

176. Luria, Introductory Comments to Jesup Lectures, in Jesup Lectures Folder, SEL Papers, APS. These lectures were not published, and I have been unable to find any correspondence about why they were not considered for inclusion in the Columbia Biological Series.

177. Luria, Jesup Lecture I, p. 1, Jesup Lectures Folder, SEL Papers, APS.

178. Luria, Jesup Lecture I, p. 1, Jesup Lectures Folder, SEL Papers, APS.

179. Luria, Jesup Lecture I, pp. 2–3, Jesup Lectures Folder, SEL Papers, APS.

180. Luria, Jesup Lecture I, pp. 3, 5–6, Lecture III, p. 2, Jesup Lectures Folder, SEL Papers, APS.

181. Luria, Jesup Lecture VI, Jesup Lectures Folder, SEL Papers, APS.

182. Luria, Jesup Lecture V, pp. 1–2, Jesup Lectures Folder, SEL Papers, APS.

183. Luria, Jesup Lecture VII, outline notes p. 2, Jesup Lectures Folder, SEL Papers, APS.

184. Luria, Jesup Lecture II, p. 7, Jesup Lectures Folder, SEL Papers, APS.

185. Luria, Jesup Lecture III, p.1, Jesup Lectures Folder, SEL Papers, APS.

186. Luria, "Bacteriophage: An Essay on Virus Reproduction," in Delbrück, *Viruses 1950*, with an addendum on "The Distribution of Spontaneous Phage Mutants from Individual Bacteria"; Luria, "Bacteriophage: An Essay of Virus Reproduction," *Science* 111 (May 12, 1950): 507–511.

187. Luria, "Bacteriophage," *Science* 111 (May 12, 1950): 511.

188. Luria, "Bacteriophage," *Science* 111 (May 12, 1950): 511.

189. Luria, *General Virology*, ix.

190. Taped interview with L. S. McClung, May 9, 1968, American Society for Microbiology Archives, Baltimore, MD.

191. Luria, *General Virology*, x–xi.

192. Hughes, *Virus: A History*, 104–105.

193. Correspondence between Luria and Demerec, February 3 to March 1, 1947, Salvador Luria Folder, Carnegie Institute of Washington Correspondence, Cold Spring Harbor Laboratory Archives, Cold Spring Harbor, NY. Hereafter CSHL.

194. Letter to Dean Ashton from L. S. McClung, July 30, 1948, Salvador Edward Luria Folder, Wells Papers.

195. Letter to Dean Ashton from Ralph Cleland, July 28, 1948, Salvador Edward Luria Folder, Wells Papers.

196. He doesn't say whether this is a personality issue or because of his political activity.

197. Letter from Muller to Wells, January 28, 1950, Luria Folder, Muller Papers, Lily Library, Indiana University.

198. Letters from Sonneborn and Muller to H. O. Halvorson, March 14 and 15, 1950, in S. E. Luria Staff Appointment File (Record Series 2/5/15), University of Illinois Urbana-Champaign Archives, Urbana, IL. Hereafter SEL UI Personnel File.

199. Letter from Sonneborn, March 14, 1950, SEL UI Personnel File.

200. Letter from Halvorson to Dean Ridenour, March 17, 1950, SEL UI Personnel File; "I.U. Accepts Resignations of Two Professors," *Indiana Daily Student*, June 23, 1950, in Luria Clippings File, Indiana University Archives.

201. Handwritten note dated April 7, 1972, Muller Papers. There is nothing in the minutes of the trustees meeting about what people thought of Luria, just that they accepted his resignation. Luria also suspected that this was the reason that they let him go. See Luria, *Slot Machine*, 43.

FREEDOM FOR SCIENCE IN COLD WAR AMERICA

1. "Message to SAC, Indianapolis" to Communications Section from Director, FBI, October 24, 1950, FBI-SEL. Similar to the Chambers example, it is interesting to note that Pontecorvo's name is not blacked out to protect his privacy, although the names of other Communist associates were. Apparently, defectors lose their right to privacy when they leave the United States.

2. See chapter 1 for a description of Luria's time with Fermi.

3. While he was in the United States, the FBI office in Oklahoma had watched Pontecorvo carefully under the directive of "Internal Security—R, Alien Enemy Control," and they had found publications about Communism in his home. See Memorandum to SAC, Indianapolis, from Director, FBI, October 26, 1950, FBI-SEL about changing Luria's investigation code.

4. See entries in Bilensky et al., *B. Pontecorvo Selected Scientific Works*; and Mafai, *Il Lungo Freddo*.

5. Davies, "Soviet Atomic Espionage," 143–144. Davies lists Pontecorvo right after convicted spy Klaus Fuchs, and he claims "circumstances of his flight make this participation [in spying activity] very likely."

6. "Memorandum" to SAC, Indianapolis, from Director, FBI, October 26, 1950, p. 2, FBI-SEL. Springfield, Illinois, had been made the office of origin in the case in a letter dated October 25, but the change apparently had not yet taken effect the next day.

7. "Purpose" and "Background" of Memorandum from C. E. Hennrich to A. H. Belmont, November 25, 1950, p. 1, FBI-SEL. I am not sure who these two individuals are, but they seem to be in the Washington office.

8. "Background" of Memorandum from C. E. Hennrich to A. H. Belmont, November 25, 1950, p. 2, FBI-SEL.

9. "Observation" of Memorandum from C. E. Hennrich to A. H. Belmont, November 25, 1950, p. 2, FBI-SEL.

10. "Report" from Springfield SAC, December 7, 1950, p. 2, FBI-SEL.

11. "Report" from Springfield SAC, December 7, 1950, pp. 2–3, FBI-SEL.

12. "Report" from Springfield SAC, December 7, 1950, p. 5, FBI-SEL.

13. "Administrative Details" in "Report" from Springfield SAC, December 7, 1950, p. 6, FBI-SEL.

14. "Administrative Details" in "Report" from Springfield SAC, December 7, 1950, p. 6, FBI-SEL.

15. See "Report" from Chicago, December 29, 1950, p. 2, FBI-SEL, and other reports from Chicago, Seattle, and Washington.

16. Fano's name is blacked out in the report, of course, but the censor slipped on another document. On a memorandum from the FBI director to the SAC in New York, the director referred to the WFO (Washington Field Office) report on an interview with man a who could not link Luria to the Communist Party. The director suggests reinterviewing a different female informant "concerning the statements which she attributed to Fano." These statements, which appeared in the New York report from July 1950 (as described above), were that Luria had been a Communist Party member in Italy, and that he had been ordered to return to Italy and resume CP activity there after his arrival in the United States. "Memorandum" to SAC, New York, from Director, FBI, January 27, 1951, FBI-SEL. Biographical information about the individual interviewed by the Washington Field Office also points to Fano: a friend from Turin, educated with him, who left Italy in 1938, and who had seen Luria several times a year.

17. "At Washington, D.C." Report from Washington Field Office, January 25, 1951, p. 3, FBI-SEL.

18. "At Washington, D.C." Report from Washington Field Office, January 25, 1951, p. 3, FBI-SEL.

19. "Memorandum" from SAC, New York, to Director, FBI, February 9, 1951, FBI-SEL.

20. "Ltr to Bureau" from New York, January 6, 1951, FBI-SEL.

21. Report from Springfield, May 24, 1951, p. 2, FBI-SEL.

22. Constitution of the University of Illinois American Federation of Teachers, Local 1055, University of Illinois Subject File, Series IIa, SEL Papers, APS; "Local FAS Chapter Asks Reversal on Oppenheimer," *Daily Illini*, June 5, 1954, in Academic Freedom Folder, George Stoddard Papers, University of Illinois at Urbana-Champaign Archives, Series 2/10/1, Box 68. Hereafter Stoddard Papers.

23. Schrecker, *No Ivory Tower*, 12.

24. "Freedom of Inquiry: A Statement of Principle," sent at the request of R. Will Burnett, April 12, 1951, University of Illinois Correspondence Folder, Series I, SEL Papers, APS. I am not sure if Luria helped write the statement, but I am pretty confident that he was at the meeting.

25. "Intellectual Freedom: A Proposed Statement of Principle," pp. 1–2, n.d., University of Illinois Correspondence Folder, Series I, SEL Papers, APS.

26. "Intellectual Freedom: A Proposed Statement of Principle" p. 2, n.d., University of Illinois Correspondence Folder, Series I, SEL Papers, APS.

27. "Dear Colleague" letter, May 23, 1951, signed by Luria and others, University of Illinois Correspondence Folder, Series I, SEL Papers, APS.

28. Letter to the Informal Committee on Intellectual Freedom and "Dear Colleague" Letter, both from B. Othanel Smith, November 5 and 15, 1951, University of Illinois Correspondence Folder, Series I, SEL Papers, APS.

29. Report from Springfield, October 18, 1951, pp. 1–3, FBI-SEL.

30. Report from Springfield, January 11, 1952, p. 2, FBI-SEL.

31. Report from Springfield, January 11, 1952, p. 1, FBI-SEL.

32. "Memorandum" from J. E. McMahon to W. A. Branigan, June 3, 1952, p. 2, FBI-SEL.

33. Letter from SEL to Passport Division, December 28, 1951, U.S. Department of State, Passport Division Folder, SEL Papers. These documents all come from Luria's correspondence: the U.S. Department of State records are extremely difficult to track down, and despite my best attempts to get them through a FOIA request and a search of the National Archives, I cannot find them. This correspondence is reported, almost verbatim, in Luria's FBI file, but there is no additional documentation from the State Department.

34. "News and Notes: Scientists in the News," *Science* 115 (January 18, 1952): 54.

35. "Comments and Communications," *Science* 116 (July 25, 1952): 95–96; Kutler, *American Inquisition*, 89–117, at 97; Caute, *The Great Fear,* 245–248; Schrecker, *No Ivory Tower*, 296–297. Shipley's investigations seem to have been in tandem with the

FBI's—she does not seem to have requested files from the FBI but investigated individuals and groups on her own. Kutler, *American Inquisition*, 93.

36. Quoted in Kutler, *American Inquisition*, 93.

37. Kutler, *American Inquisition*, 92. Comments also published in *Science* 116 (August 29, 1952): 234–236.

38. Letter from Ruth Shipley, U.S. Department of State, to SEL, January 25, 1952. U.S. Department of State, Passport Division Folder, SEL Papers, APS.

39. Letter from SEL to R. B. Shipley, January 29, 1952, p. 1. U.S. Department of State, Passport Division Folder, SEL Papers, APS.

40. Letter from SEL to R. B. Shipley, January 29, 1952, pp. 1–2. U.S. Department of State, Passport Division Folder, SEL Papers, APS.

41. Letter from SEL to R. B. Shipley, January 29, 1952, p. 2. U.S. Department of State, Passport Division Folder, SEL Papers, APS.

42. Letter from R. B. Shipley to SEL, February 15, 1952. U.S. Department of State, Passport Division Folder, SEL Papers, APS.

43. According to Zella Luria, he did not want to become a court case. Author interview with Zella Hurwitz Luria, January 13, 2000. Luria's letter to the society is not in his papers, but the response from the organization is. Jonathan [surname unknown] wrote back, saying, "I am both surprised and pained that you have been refused a passport. . . . Another unfortunate blow to Anglo-American relations." He then reassured Luria that the society would support him in whatever decision he made, and that it would be done discreetly. Letter in Society for General Microbiology Folder, February 4, 1952, SEL Papers, APS.

44. Letter from George Stoddard to Howland H. Sargeant, March 1, 1952, p. 1, Salvatore E. Luria Staff Appointment File (Record Series 2/5/15), University of Illinois at Urbana-Champaign Archives. Hereafter SEL Personnel File.

45. The oath that Luria signed is probably no longer in existence. Every employee was required to sign another loyalty oath in 1955, which was filed in the bursar's office, but there are no records of these oaths at the University of Illinois archives. The blank copy of the oath can be found in "Loyalty Oath" Folder, Record Series number 48/1/5, Box 7, University of Illinois at Urbana-Champaign Archives.

46. Letter from George Stoddard to Howland H. Sargeant, March 1, 1952, p. 2, SEL Personnel File.

47. Somehow, the FBI got a copy of Luria's correspondence with Shipley through an informant at the University of Illinois. Report from Springfield, May 6, 1952, SEL-FBI.

48. Letter from Howland H. Sargeant to George Stoddard, March 12, 1952, SEL Personnel File.

49. Letter from George Stoddard to Howland H. Sargeant, March 27, 1952, SEL Personnel File.

50. Letter from George Stoddard to Howland H. Sargeant, March 27, 1952, and photocopy of "University Scientists' Work Helps Effort to Control Cancer Virus," *Daily Illini*, February 19, 1952, SEL Personnel File.

51. Letters and Secret Air Courier from Hoover, March 19, 1952, SEL-FBI.

52. "Memorandum" from Springfield SAC to Director, FBI, March 12, 1952. The Washington report from May 22 does not contain any additional information from the State Department files.

53. Letter from SEL to James Watson, April 10, 1952, Luria Correspondence, James D. Watson Papers, Cold Spring Harbor Laboratory Archives, Cold Spring Harbor, NY.

54. Watson, *Double Helix*, 71–72. Watson's comment on Luria's passport refusal was "As usual, the State Department would not come clean on what it considered dirt."

55. Stahl, "Hershey" in *We Can Sleep Later*, 3.

56. On Pauling's encounter with Ruth Shipley, see "Summary of Testimony of Linus Pauling," 28; Watson, *Double Helix*, 71; Caute, *Great Fear*, 471–473; Kutler, *American Inquisition*, 89–91; Goertzel and Ben Goertzel, *Linus Pauling*, 122–125; Hager, *Force of Nature*, 400–407. Luria and Pauling were not the only scientists who were denied passports. See letters published in *Science* from Paul Erdos, Ernest Borek and Frances Ryan, Jacques Monod, and the Florida Committee on Science and Public Affairs regarding passport and visa refusals. *Science* 116 (August 15, 1952): 178–179. See also *Bulletin of the Atomic Scientists* from October 1952, which was a special issue devoted to visa problems for foreign scientists.

57. Watson, *Double Helix*, 80–81, 91–95, 99; Kutler, *American Inquisition*, 89–91.

58. Antoinette Pirie to SEL, June 21, 1952, Pirie Correspondence Folder, Series I, SEL Papers APS. She wrote, "At one of these [open sessions of the meeting] a member asked, "in view of the rumors that Prof. Luria was prevented from attending the meeting could the Sec. deny these rumors. The reply was given that the Sec. was not in a position to deny these rumors. It is certain that at the meeting, . . . the rumors were assumed to be correct." On the lack of public comment at the meeting, see *The Nature of Virus Multiplication: The Second Society for General Microbiology Meeting Held at Oxford University, April 1952* (Cambridge: Published for the Society for General Microbiology by the University Press, 1953).

59. "Memorandum" to Director, FBI from SAC, Washington, March 27, 1959, FBI-SEL.

60. Bertani, "Lysogeny at Mid-Twentieth Century."

61. Luria, "Bacteriophage," *Science* 111 (May 12, 1950): 509.

62. Luria, *Slot Machine*, 98.

63. Luria, "T2 Mystery," 92.

64. Luria and Human, "Nonhereditary, Host-Induced Variation," 557.

65. Luria, *Slot Machine*, 98–99. This is the "broken test tube" Luria refers to in the title of his autobiography, as the second of his major contributions to modern biology. In "The T2 Mystery," he told the story without the broken test tube and gave Human credit for trying the experiment with *Shigella* instead.

66. Luria and Human, "Host-Induced Variation," 557.

67. Luria, "Host-Induced Modification of Viruses," 237.

68. Luria, "T2 Mystery," 93; Luria, *Slot Machine*, 99.

69. Delbrück, "Introductory Remarks about the Program."

70. "1953 Symposium" in Inglis, Sambrook, and Witkowski, *Inspiring Science*, pp. 107–136, particularly "Memories of the 1953 Symposium," pp. 110–115.

71. Luria, "Host-Induced Modifications of Viruses," 237–244.

72. Luria cites Bertani, Wiegle, Ralson, and Krueger, who had made similar observations around the same time with other phages and bacterial hosts. Luria, "Host-Induced Modification of Viruses," 238.

73. Luria, "Plan for Research," October 7, 1952, John Simon Guggenheim Memorial Foundation Folder, SEL Papers, APS.

74. Luria, "Host-Induced Modification of Viruses," 241–242.

75. Luria, "Host-Induced Modification of Viruses," 244; Luria also discusses the analogy to animal diseases, including cancer, in "The T2 Mystery," 98.

76. Luria, "T2 Mystery," 98.

77. Brock, *Emergence of Bacterial Genetics*, 325–326.

78. Sidik, "The Untidy Experiment."

79. Luria, "Ethical and Institutional Aspects," 57; Luria, *Slot Machine*, 101.

80. Henrietta De Boer, "Paper on the History of the 1953 Fight Against the Broyles Bills, 1954–5," pp. 1–2, in John J. De Boer Papers, University of Illinois at Urbana-Champaign Archives Series10/7/20, Box 2. Hereafter Boer Papers. This is an informal document based on news accounts and her experience as the secretary of the Champaign-Urbana Committee to Oppose the Broyles Bills. Unfortunately, her list of footnotes is not in the file, just the document. See also a brief mention in Schrecker, *No Ivory Tower*, 112–113.

81. Schrecker, *No Ivory Tower*, 194.

82. De Boer, "Paper on the History of the 1953 Fight Against the Broyles Bills," 4.

83. On the California oath, see Gardner, *California Oath Controversy*; Schrecker, *No Ivory Tower*, 117–125; and Kaiser, "Nuclear Democracy." I am grateful to David Kaiser for pointing this distinction out to me.

84. Description in De Boer, "Paper on the History of the 1953 Fight," and "Resolution Adopted by the University of Illinois Senate on Senate Bills 101 and 102 Pending Before the 68th General Assembly of the State of Illinois," in Academic Freedom Folder, Stoddard Papers, Box 68.

85. De Boer, "Paper on the History of the 1953 Fight," 4.

86. De Boer, "Paper on the History of the 1953 Fight," 5–7.

87. "Psychologists Fight Broyles Bills; Rusk Modifies Stand," *News Gazette* (Urbana, IL), May 17, 1953. Zella Luria was one of the signers of a statement against the bills. "Dear Colleague" letter from A Committee of Mathematicians Against the Broyles Bills, April 16, 1953, in Duplicated Materials for Distribution and Newspaper Clippings 1953 Folder, Box 2, De Boer Papers.

88. "Resolution Adopted by the University of Illinois Senate," 3.

89. "Resolution Adopted by the University of Illinois Senate," 5–6.

90. "14 Controversies that Marked Stoddard's Administration," p. 5, n.d., in University of Illinois Subject Folder, Series IIb, SEL Papers, APS.

91. "14 Controversies that Marked Stoddard's Administration," p. 4; also mentioned in Stoddard, *Pursuit of Education*, 129–130.

92. Quoted in De Boer, "Paper on the History of the 1953 Fight," 28.

93. Quoted in De Boer, "Paper on the History of the 1953 Fight," 28–29.

94. Schrecker, *No Ivory Tower*, 258.

95. Stoddard, *In Pursuit of Education*, 127–138.

96. Letter from Luria to Gunsalus, July 27, 1953, Luria Correspondence Folder, Box 29, Irwin Gunsalus Papers, University of Illinois at Urbana-Champaign Archives Series 15/5/40. Hereafter Gunsalus Papers.

97. Luria reassured Stoddard that the political climate would not interfere with his scientific research. "You may be sure that we shall consider it an assignment still to be carried out, unless a time should come when remaining here would mean accepting and subscribing to the evils that led to your forced resignation." Letter from Luria to Stoddard, September 15, 1953, and Stoddard's response, September 25, 1953, in University of Illinois Correspondence Folder, Series I, SEL Papers, APS.

98. I didn't see anything in the Illinois files about Cazden's case.

99. Quoted in De Boer, "Paper on the History of the 1953 Fight," 34.

100. Invitations to 1st Meeting, January 31, 1955, and Minutes from the February 4 meeting, in Minutes January-June, Box 3, De Boer Papers.

101. On Lederberg, see Academic Freedom-Loyalty Procedure Folder, Box 12, Lloyd Morey Papers, University of Illinois at Urbana-Champaign Archive Series 2/11/1. Hereafter Morey Papers. "Committees and Liaisons," in Action 1955 Folder, Box 3, De Boer Papers.

102. "Ask UI Faculty to Join Civil Liberties Union," *News-Gazette*, April 1955 (no date on clipping), Action 1955 Folder, Box 3, De Boer Papers.

103. Public meeting planning notes, April 1955, Action 1955 Folder, Box 3, De Boer Papers.

104. "Senate Opposes Broyles Bills: Fifth Amendment Rules set at U.I.," *Daily Illini*, April 6, 1955.

105. On July 7, Gunsalus wrote to Luria, "The Legislature closed but there is very little in the papers as to what has been and will be signed. Very possibly the issue is dead. The people from here did contact the Governor, which as I understand it was somewhat effective before." Letter from Gunsalus to Luria, July 7, 1955, Luria Correspondence Folder, Box 29, Gunsalus Papers.

106. Memorandum to the Board of Trustees, July 20, 1955, in Academic Freedom-Loyalty Procedure Folder, Box 12, Morey Papers.

107. The bursar's office presumably kept these on file, but the holdings in the university archives don't go back that far. Only a handful of professors refused to sign, and Luria was not one of them.

108. On Luria's election, see "Association Affairs," *Science* 121 (April 1, 1955): 475. On the organization of the AAAS Council, see Wolfle, *Renewing a Scientific Society*, 5.

109. "Preliminary Announcement of Atlanta Meeting," *Science* 121 (May 20, 1955): 751.

110. On the Section H cancellation, see Angel, "Physical Anthropologists." On Cobb, see Rankin-Hill and Blakey, "W. Montague Cobb."

111. Board of Directors, American Association for the Advancement of Science, "Advantages of an Atlanta Meeting," *Science* 121 (June 24, 1955): 7A.

112. "Hotel Headquarters and Housing, Atlanta Meeting" in "Scientific Meetings," *Science* 122 (July 22, 1955): 165. The published preview and the report afterward do not mention that segregation had been a concern at all. See Raymond L. Taylor, "Preview of AAAS Meeting, Atlanta," *Science* 122 (December 2, 1955): 1061–1081; and Raymond L. Taylor, "Atlanta Meeting in Retrospect," *Science* 123 (February 17, 1956): 273–279. The report of the council meeting does discuss segregation.

113. Letter from SEL to W. Montague Cobb, August 14, 1955, AAAS Folder, SEL Papers, APS. Luria was a member of the NAACP in 1953—it is unclear how long he was a

member of that organization. "National Association for the Advancement of Colored People" Folder, Series IIa, SEL Papers, APS.

114. Copies of Luria's letters are not in his archive, but Wolfle's tepid response directing Luria to the June Statement in Science is. Luria and Weaver seem to have discussed the matter on the phone in July, according to a letter Luria wrote to Weaver in October. Letter from Wolfle to SEL, August 19, 1955, and letter to Weaver from SEL, October 5, 1955, AAAS Folder, SEL Papers, APS.

115. Letter from Cobb to SEL, August 20, 1955, AAAS Folder, SEL Papers, APS.

116. These were some of the larger sections of the AAAS. Letters from Gabriel Lasker to SEL, Harold Plough and Barry Commoner, September 23, 1955, AAAS Folder, SEL Papers, APS.

117. Proposed Resolutions in Letter from Gabriel Lasker to SEL, etc., September 23, 1955, AAAS Folder, SEL Papers, APS.

118. Letter from SEL to Lasker, September 27, 1955, AAAS Folder, SEL Papers, APS.

119. Letter to George Beadle from SEL, October 5, 1955, AAAS Folder, SEL Papers, APS. The letter to Weaver on the same date is shorter but contains similar language.

120. Letter to SEL from Beadle, October 7, 1955, AAAS Folder, SEL Papers, APS.

121. Letter to SEL from Weaver, October 14, 1955, AAAS Folder, SEL Papers, APS.

122. Letter to Quentin D. Young from SEL, October 17, 1955, AAAS Folder, SEL Papers, APS.

123. Letters to Luria and Beadle from C. G. Van Arman, secretary of the Illinois Section of the Society for Experimental Biology and Medicine, December 7, 1955, AAAS Folder, SEL Papers, APS.

124. Letter to Dr. Max M. Friedman from Harold Feinberg, Secretary Chicago Section American Association of Clinical Chemists (cc to Luria), December 16, 1955, AAAS Folder, SEL Papers, APS.

125. Wolfle, "AAAS Council Meeting, 1955," 268.

126. "Resolution—AAAS," letter to Dael Wolfle and council members from Bernard Davis, December 22, 1955, AAAS Folder, SEL Papers, APS.

127. Wolfle, "AAAS Council Meeting," 268. Cover letter to council members from Wolfle, December 30, 1955, AAAS Folder, SEL Papers, APS.

128. Letter to Cobb from SEL, January 24, 1956, AAAS Folder, SEL Papers, APS.

129. Wolfle, *Renewing a Scientific Society*, 55.

130. Schoenfeld, "Anti-Nepotism Rules"; Rossiter, *Women Scientists in America*, chap 6. Interestingly, the issue does not appear at all in an American Association of

University Professors study of the effects of the Depression on higher education from 1937. See Willey, *Depression, Recovery and Higher Education.*

131. Dolan and Davis, "Antinepotism Rules," 286.

132. Rossiter shows that the Illinois psychology department had very few women to begin with—only one full professor and three additional women at the assistant or associate level—although she does not say how many faculty members there were in total. See tables 6.2 and 6.3 in Rossiter, *Women Scientists*, 132–136.

133. Memo from A. C. Ivy to George Stoddard, September 21, 1946, and "Policy on Nepotism" with revisions in 1947 and 1952, September 3, 1953, in Nepotism P. F. Folder, Box 18, Lloyd Morey Papers, Series 2/11/1, University of Illinois Archives, Urbana, IL; Rossiter, *Women Scientists*, 125.

134. Author interview with Zella Hurwitz Luria, January 15, 2000.

135. University of Wisconsin Folder, Series I, and Albert Einstein School of Medicine Folder. In addition, there is a reference to an offer from UCLA in a letter from Cyrus Levinthal, January 13, 1958, Cyrus Levinthal Folder, Series I, SEL Papers, APS.

136. Letter from SEL to Max Delbrück, December 12, 1958, Luria Correspondence, Max Delbrück Papers.

137. Letter from Dean Wall to Zella Luria, February 11, 1959, and from Dean Lanier to Salvador Luria, February 13, 1959, University of Illinois Folder, Series I, SEL Papers, APS.

138. Letter from SEL to Dean Lanier, February 19, 1959, University of Illinois Folder, Series I, SEL Papers, APS.

RECOGNITION AND RESPONSIBILITY

1. On the history of practicality in American science, see, for example, Graham, "The Necessity of the Tension."

2. Annual Letter to Alumni and Friends of the Department of Biology, Massachusetts Institute of Technology, Volume 5, June 1957. See Letter from Dean Stratton to Warren Weaver, September 12, 1956, and letters to the committee members, November 1956, Biology Department Folder, James Killian Presidential Papers, AC 4, Institute Archives and Special Collections, MIT Libraries, Cambridge, MA. Hereafter MIT Archives. Author interview with Boris Magasanik, April 23, 1999.

3. Rasmussen discusses this meeting in the context of the history of the department of biology in "Mid-Century Biophysics Bubble," 272–277. The other members of the committee were Paul Weiss, Hudson Hoagland, Carroll Williams, our own Salvador Luria, Walter Wilson, Glen King, and David Goddard, who represented a wide range of biological disciplines and institutions. On Warren Weaver's long-standing interest

in and influence on the development of molecular biology, see Kohler, "Management of Science"; Abir-Am, "Discourse of Physical Power"; and Kay, *Molecular Vision of Life*.

4. Rasmussen, "Mid-Century Biophysics Bubble," 273.

5. "The Future of Biology at MIT" Section of Annual Letter 1957, 5.

6. Magasanik, "Charmed Life," 8.

7. Botanists seem to have been hit particularly hard by this trend. See Bonner, "Future Welfare of Botany"; Laetsch, "Welfare of Botany Re-Examined"; and Shetler, "Botany—A Passing Phase?." Smocovitis briefly discusses this perceived crisis at the end of *Unifying Biology*, 172–178.

8. Grant and Saul, "Curricular 'Straw Man' for Biologists," 15.

9. Grobman, "BSCS: A Challenge," 20.

10. Rudolph, *Scientists in the Classroom*, Chapter 6; Wolfe, "Biology and Liberty for All"; Wolfe, *Freedom's Laboratory*, chap. 7. These sources focus on high school, rather than college or graduate education.

11. Bonner, "Editorial: Thoughts About Biology."

12. Grant and Saul, "Curricular 'Straw Man,'" 15.

13. Appel, *Shaping Biology*, 207.

14. Sizer, "Future of the Life Sciences," 2, 4–7.

15. Memo from Sizer to Killian, October 31, 1964, Biology Department Folder, Killian Presidential Papers, AC 134, MIT Archives.

16. From Letter of Recommendation from Max Delbrück to Irwin Sizer, January 9, 1959, Luria correspondence, Delbrück Papers.

17. Sizer, "Future of the Life Sciences," 15. Luria's name appears on page 13. I was unable to see the list because of archive policy restricting access to personnel information, but the archivist confirmed that Luria's name is on the list.

18. Letter from Victor Weisskopf to J. A. Stratton, December 12, 1958, Biology Department Folder 3, Stratton Papers, AC 134, MIT Archives. There is nothing in the Delbrück-Weisskopf correspondence about Luria, although Delbrück did send Irwin Sizer "an appreciation of Dr. Luria's qualifications." Letter from Delbrück to Sizer, January 9, 1959, Luria correspondence, Delbrück Papers.

19. Letter from Victor Weisskopf to J. A. Stratton, December 16, 1958, Biology Department Folder 3, Stratton Papers, AC 134, MIT Archives.

20. Letter from Weisskopf to Stratton, December 16, 1958; Harrison and Sizer to Stratton, p. 1, Biology Department Folder 3, Stratton Papers, AC 134, MIT Archives.

21. Letter from SEL to Dean Lanier, February 19, 1959, University of Illinois Folder, Series I, SEL Papers, APS; taped interview with L. S. McClung, May 9, 1968, American Society for Microbiology Archives, Baltimore, MD.

22. After a semester at MIT, Levinthal wrote to Luria, "Obviously things are not as glorious here as I had hoped last spring, but they are in fact a remarkably good approximation. . . . It's a little embarrassing to be able to get things you ask for and then not have any excuses left when nothing comes out of it in the end." Letter from Levinthal to Luria, January 1958, Levinthal Correspondence Folder, Series I, SEL Papers, APS.

23. Memo from Sizer to Stratton, February 15, 1959, Biology Department Folder 3, Stratton Papers, AC 134, MIT Archives.

24. Memo from Sizer to Harrison, February 24, 1959, p. 1, Biology Department Folder 3, Stratton Papers, AC 134, MIT Archives.

25. Memo from Sizer to Harrison, February 24, 1959, p. 2, Biology Department Folder 3, Stratton Papers, AC 134, MIT Archives.

26. Author interview with Nancy Ahlquist, April 20, 1999.

27. Luria, *Slot Machine*, pp. 48–49.

28. Author interview with Nancy Ahlquist, April 20, 1999.

29. Memo from Sizer to Stratton, February 15, 1959, Biology Department Folder 3, Stratton Papers, AC 134, MIT Archives.

30. Magasanik, "Charmed Life," p. 8.

31. "Report of the President, 1959," pp. 29, 182–185, Box 1, Stratton Papers, AC 134, MIT Archives.

32. "Suggestions on points Dr. Luria might consider mentioning in his presentation." Attached to letter to Luria from Harry M. Weaver, vice president for research, February 10, 1959, American Cancer Society Folder, Series I, SEL Papers, APS.

33. Letter to Luria from Harry M. Weaver, vice president for research, February 10, 1959, p. 2, American Cancer Society Folder, Series I, SEL Papers, APS. See also Gaudillière, "Molecularization of Cancer Etiology," 159–161.

34. James Watson, "Molecular Biological Approach to the Cancer Problem," presented on November 28, 1970, at a meeting on the Social Impact of Modern Biology, sponsored by the British Society for Social Responsibility in Science. MS in AC 8, Box 53, Cancer Research Folder, MIT Archives.

35. Ross, *Crusade*, 122–123. See also Feffer, "Atoms, Cancer and Politics," for a discussion of early attempts to link other types of basic research to cancer, and Gaudillière, "Molecularization of Cancer Etiology," 151–152.

36. See, for example, *Indiana Alumni Magazine*, "Lecturer Abroad," 7, and a letter Luria wrote to Patrick McGrady on December 10, 1956, saying, "There are two items in our work that may make good stories." American Cancer Society Folder, Series I, SEL Papers, APS.

37. Draft and letter to Luria, January 29, 1952, American Cancer Society Folder, Series I, SEL Papers, APS.

38. Letter from Luria to Warren H. Cole, March 26, 1952, American Cancer Society Folder, Series I, SEL Papers, APS.

39. Luria, "Viruses, Cancer Cells," 677.

40. Luria, "Viruses, Cancer Cells," 679–680.

41. Luria, "Viruses, Cancer Cells," 686.

42. Luria, "Viruses, Cancer Cells," 680–684.

43. C. H. Andrewes, in particular, was quite vehement in his disagreements. In his "Discussion" comments, he notes, "No doubt . . . he [Luria] is trailing his coat in the hope of having some sharp arguments. Very well!" See Andrewes, "Discussion of Dr. Luria's Paper," 690. On the other hand, Francois Jacob and Joseph W. Beard accepted Luria's arguments and offered other evidence to support his claims. See F. Jacob, "Comments" and Joseph W. Beard, "Comments," *Cancer Research* 20 (June 1960): 695–697, 704–705.

44. Luria, "Viruses, Cancer Cells," 678.

45. On infective cellular heredity, Andrewes wrote, "Generally accepted? This idea may or may not be a good one, but if he really thinks it is generally accepted it just convinces me that these geneticists never come out of their ivory towers. We are, of course, not ordinary virologists here, but if he went to a gathering of ordinary virologists he not only would not find it was generally accepted but that they would have great difficulty in understanding what he was talking about." Andrewes, "Discussion of Dr. Luria's Paper," 690; Gaudillière, "Molecularization of Cancer Etiology," 159–160.

46. Gaudillière, "Molecularization of Cancer Etiology," 161; Scheffler, *Contagious Cause,* 57–59.

47. "Program Proposal," draft in MIT Folder, Series IIa, SEL Papers, APS.

48. "Annual Letter to Alumni and Friends of the Department of Biology, Massachusetts Institute of Technology," nos. 8–10, 1960–1962, in Biology Department Folder, Stratton Papers, AC 134, MIT Archives.

49. Press Release, January 18, 1960, Biology Department Folder 3, Stratton Papers, AC 134, MIT Archives.

50. Press Release, January 18, 1960, Biology Department Folder 3, Stratton Papers, AC 134, MIT Archives. See also *New York Times*, "Study of Life's Origin."

51. American Academy of Arts and Sciences, "Salvador Edward Luria," https://www .amacad.org/person/salvador-edward-luria; *New York Times*, "Science Academy Adds 35 to Rolls."

52. Crotty, *Ahead of the Curve*, 31; Massachusetts Institute of Technology, "Annual Letter to Alumni and Friends," 3.

53. Author interview with James Darnell, July 2, 2001.

54. See, for example, Holland, "Review of *Virology*"; and Franklin, "Virology after 14 Years of Discovery."

55. "Department of Biology" listing in MIT Bulletin for the academic years 1957– 1958 and 1964–1965, Volume 92, July 1957, and Volume 99, July 1964; author interview with Maurice Fox, July 24, 2001.

56. MIT, "Annual Letter to Alumni and Friends," 2–3.

57. Author interview with Nancy Ahlquist, April 20, 1999.

58. "Undergraduate Program: Quantitative Biology" description, MIT Bulletin for 1957–1958, 153.

59. See course catalogs in the MIT Bulletin 1959–1970.

60. Luria, "On Teaching Biology," 131. He expressed his excitement about this new teaching job to Leland McClung in their May 9, 1968, interview. Interview in American Society for Microbiology Archives, Baltimore, MD.

61. Luria, *36 Lectures in Biology*, xv.

62. Luria, *36 Lectures*, xv.

63. Luria, "On Teaching," 131.

64. Luria, "On Teaching," 134.

65. Luria, *36 Lectures*, 358–359.

66. Luria, *36 Lectures*, 359.

67. Luria, *Slot Machine*, 128–129.

68. Luria, "Growing Up Sensible" Convocation Address at Brown University Commencement, 1975, p. 6. SEL Papers, APS.

69. Luria, *Slot Machine*, 150.

70. Luria, "Reading Lists and Notes for Literary Seminar, 1966–1969," SEL Papers, APS; Luria, "Growing Up Sensible," 6.

71. "The Life Sciences and the Humanities," July 17, 1969, and "Council for Biology in Human Affairs" July 30, 1969, Salk Institute Correspondence Folder, SEL Papers, APS.

72. Wade, "Salk Institute"; Luria, *Slot Machine*, 56–57; Jacobs, *Jonas Salk*, 263–264.

73. MIT Technology and Culture Seminar Folders, Series I, SEL Papers, APS; Keith-Lucas, "Why Radius?"

74. Several documents in the Boston Area Faculty Group on Public Issues (BAFGOPI) folder in the Luria Collection at the APS refer to statements made by Luria, Meselson, Feld, and others in the *Saturday Evening Post* arguing against testing, and the group's first public campaign was an open letter to President Kennedy about civil defense, in the *New York Times*, November 10, 1961.

75. BAFGOPI membership list in the BAFGOPI folder, Series IIa, SEL Papers, APS.

76. Alan Matusow emphasizes the role of "elite intellectuals" from New York City and Cambridge in electing and advising Kennedy and Johnson in *The Unraveling of America: A History of Liberalism in the 1960s*, especially in chap. 1.

77. "Dear Colleague" letter from BAFGOPI steering committee, p. 1, 1962, BAFGOPI folder, SEL Papers, APS. Also in BAFGOPI folder, Everett Mendelsohn Papers, Harvard University Archives, Cambridge, MA. Hereafter Mendelsohn Papers.

78. Selections from Teller, *Legacy of Hiroshima*, were excerpted in the *Saturday Evening Post*, February 1962.

79. Moore, *Disrupting Science*, p. 5 and chap. 4.

80. Pauling, "Appeal by American Scientists"; Luria, *Slot Machine*, 184.

81. Orear et al., "Answer to Teller," 69–74.

82. Orear et al., "Answer to Teller," 69.

83. Orear et al., "Answer to Teller," 70.

84. Orear et al., "Answer to Teller," 70.

85. Orear et al., "Answer to Teller," 71–74.

86. Orear et al., "Answer to Teller," 71.

87. See Luria-Sonneborn correspondence, Sonneborn Papers.

88. Sonneborn, "Preface" to Sonneborn, *Control of Human Heredity*, xii, ix.

89. Luria, "Directed Genetic Change," 2–4.

90. Luria, "Directed Genetic Change," 2.

91. Luria, "Directed Genetic Change," 3.

92. Luria, "Directed Genetic Change," 3–4.

93. Luria, "Directed Genetic Change," 4.

94. Luria, "Directed Genetic Change," 16–17.

95. Luria, "Discussion of Dr. Muller's Paper," 124.

96. Letter from Luria to Delbrück, February 4, 1961, Luria Correspondence, Delbrück Papers.

97. Luria, *Slot Machine*, 109; "Plans for Research," p. 4, in an unsuccessful 1952 application for a Guggenheim Fellowship, John Simon Guggenheim Memorial Foundation Folder, Series I, SEL Papers, APS.

98. Luria, *Slot Machine*, 105; Luria, "Phage, Colicins," 391.

99. Luria, *Slot Machine*, 106.

100. Luria, "Phage, Colicins," 391.

101. Luria, *Slot Machine*, 107–110.

102. Luria, *Slot Machine*, 49–51, 144.

103. Announcement to the Faculty, October 28, 1964, and Press Release November 20, 1964, Sedgwick Professorship Folder, Killian Presidential Papers, MIT Archives.

104. Letter to SEL, June 10, 1969, Columbia University Folder, Series I, SEL Papers, APS; Luria Press Conference Transcript, October 16, 1969, p. 2, Salvador E. Luria Publicity File, MIT Museum.

105. "Louisa Gross Horwitz Prize," https://www.cuimc.columbia.edu/research/louisa-gross-horwitz-prize/history; Renzulli, "Two Biophysicists Win," http://www.columbia.edu/cu/record/archives/vol22/vol22_iss6/record2206.16.html.

106. *New York Times*, "2 Biologists Will Share."

107. Nobel Prize, https://www.nobelprize.org/prizes/medicine/1969/summary/.

108. "Summary," in the Nobel Prize Press Release, October 1969, Nobel Prize Folder, Series I, SEL Papers, APS, http://www.nobel.se/medicine/laureates/1969/press.html.

109. Stent, "1969 Nobel Prize for Physiology or Medicine," 479.

110. Baltimore, *Nobel Lectures in Molecular Biology*, viii.

111. Letter from Joshua Lederberg to SEL, October 16, 1969, Nobel Prize Folder #3 October 17 F-P, Series I, SEL Papers, APS.

112. Biographical Sketch, Dr. Salvador Edward Luria, [1969] Nobel Prize Folder #2, Series I, SEL Papers, APS.

113. Stent, "1969 Nobel Prize," 481.

114. Luria, *Slot Machine*, 54.

115. Page 4 of proofs of Presentation Speech by Professor Sven Gard, Physiology or Medicine 1969 (reproduced in Baltimore, *Nobel Lectures in Molecular Biology*), Nobel Prize Folder #2, Series I, SEL Papers, APS.

116. Luria, "Phage, Colicins," 387–388.

117. Luria, "Phage, Colicins," 396.

PROTESTING THE VIETNAM WAR

1. For general works on the 1960s, see Gitlin, *The Sixties*; Matusow, *Unraveling of America*; Farber, *The Sixties*; Farber, *Age of Great Dreams*; Brick, *Age of Contradiction*; Tomes, *Apocalypse Then*.

2. The classic history of the Vietnam conflict is Herring, *America's Longest War*. Other useful sources include Gitlin, *The Sixties*; Farber, *Age of Great Dreams*; and Buzzanco, *Vietnam and the Transformation of American Life*.

3. Many accounts of the antiwar movement, such as Gitlin's *The Sixties*, are by participants in the leadership of groups such as Students for a Democratic Society. Others evaluate the political impact of the peace movement. See DeBenedetti, "On the Significance of Citizen Peace Activism"; Katz, "Peace Liberals and Vietnam"; and Wells, *The War Within*. One early journalistic account is Zaroulis and Sullivan, *Who Spoke Up?* Recent historical analyses include DeBenedetti, *American Ordeal*; Levy, *Debate over Vietnam*; and Heineman, *Campus Wars*. The only historical analysis of professors' responses to the war is Schalk, *War and the Ivory Tower*. Some aspects of the antiwar movement by scientists and students are discussed in Moore, *Disrupting Science*.

4. BAFGOPI documents in SEL Papers, APS, and Mendelsohn Papers, Harvard.

5. See, for example, the MASSPAX statement from February 9, 1965, where Luria is listed as a member of the advisory committee, along with fellow biologists Hudson Hoagland and Albert Szent-Györgyi. Massachusetts Peace Movement Papers, Boston Public Library Special Collections.

6. On the relationship of the protest movement to existing groups, see Gitlin, *The Sixties*; Wells, *The War Within*; Zaroulis and Sullivan, *Who Spoke Up?*; and DeBenedetti, *An American Ordeal*.

7. Zaroulis and Sullivan, *Who Spoke Up?* , 34.

8. *Newsweek*, "Ad-itorial Voices."

9. Luria, *Slot Machine*, 189.

10. Luria, *Slot Machine*, 193. On Chomsky's arrest in the October 21, 1967, protest at the Pentagon, see the event chronicled in Mailer's *Armies of the Night*. See also O'Neill, *Coming Apart*; Zaroulis and Sullivan, *Who Spoke Up?*, 136–142; and Levy, *Debate over Vietnam*, 137–138.

11. *New York Times*, "Mr. President: STOP THE BOMBING."

12. Lissner, "6,000 in Colleges Sign Bomb Appeal."

13. Lissner, "6,000 in Colleges."

14. Luria, *Slot Machine*, 186.

15. Luria, "Babi Yar, Warsaw and Vietnam."

16. Luria, "Babi Yar, Warsaw and Vietnam."

17. Berenson, "Answer to Dr. Luria"; letter from Phillip Hochstein to Toby Brooks in response to her request to reprint Luria's editorial in *The Jewish Week*, April 28, 1970, Toby Brooks Folder, SEL Papers, APS.

18. Pollard, "Vietnam: Call for Scientific Help."

19. Luria and Szent-Györgyi, "Vietnam: A National Catastrophe."

20. Pollard, "Scientists' Views on Vietnam."

21. Letters from William F. Prokasy, William Palmer Taylor, and Robert B. Kelman, "Scientists' Views on Vietnam," *Science* 158 (October 27, 1967): 440–441.

22. Letter from Kirsten Emmot to SEL, November 22, 1967, Kirsten Emmot Correspondence Folder, Series I, SEL Papers, APS.

23. Positive letter from John M. Reiner, *Science* 158 (December 14, 1967): 1393.

24. Letters from Stanley Buckser and E. Staten Wynne, *Science* 158 (December 14, 1967): 1393.

25. The literature on Fort Detrick is mainly U.S. Army documentation. The main source is Covert, *Cutting Edge*. See chapters 5 and 6 for World War II projects. Also see Miller, Engelberg, and Broad, *Germs*, 39–65.

26. From the January 10, 1955, meeting of the SAB Council Policy Committee, and Council Minutes from May 8, 1955. American Society for Microbiology Archive, Baltimore. Hereafter ASM Archive.

27. "Report of the Committee Advisory to the U.S. Army Biological Laboratories," *ASM News* 33, no. 3 (August 1967): 21.

28. "Report of the Committee Advisory to the U.S. Army Biological Laboratories," *ASM News* 33, no. 3 (August 1967): 20–21.

29. "Report of the Council and Council Policy Committee 29 April—2 May, 1967," *ASM News* 33, no. 3 (August 1967): 12.

30. "Report of the Council and Council Policy Committee 29 April—2 May, 1967," *ASM News* 33, no. 3 (August 1967): 12.

31. These members are not identified by name. "Report of Annual Business Meeting, 3 May 1967," *ASM News* 33, no. 3 (August 1967): 17.

32. "Report of Annual Business Meeting, 3 May 1967," *ASM News* 33, no. 3 (August 1967): 17.

33. Letter to Raymond C. Bard from SEL, May 15, 1967, American Society for Microbiology Correspondence Folder 5, Series I, SEL Papers, APS.

34. The eighteen folders of correspondence from his year as president include a dizzying number of letters involving committee appointments and other administrative details. During his tenure, the ASM moved to a new headquarters in Washington, D.C., and formally joined its administration to that of the American Academy of Microbiology. In addition, Luria and his colleagues had to respond to several bills introduced in Congress to change the name of the *Salmonella* bacteria to something that would not sound so much like "salmon" and thus hurt the fish industry. They eventually convinced the senators from Alaska and Washington that it was inappropriate for the U.S. government to legislate scientific nomenclature. See ASM Correspondence, Series I, SEL Papers, APS.

35. Letter from Alvin J. Clark to SEL, August 10, 1967, ASM Folder 7, Series I, SEL Papers, APS.

36. Clark to SEL, August 10, 1967, 3.

37. The two main individuals were David Dubnaw and Isaar Smith in the New York branch. Clark to SEL, August 10, 1967, 4–5.

38. "Special Biological Warfare Meeting—Northern California Branch," ASM Folder 10, Series I, SEL Papers, APS.

39. Quoted in President's Message in American Society for Microbiology Newsletter Northern California Branch, November 1967, 2, ASM Folder 10, Series I, SEL Papers, APS.

40. "Special Biological Warfare Meeting—Northern California Branch," 1, ASM Folder 10, Series I, SEL Papers, APS.

41. Letter from Clark to SEL, November 27, 1967, ASM Folder 10, Series I, SEL Papers, APS.

42. Letter from James W. Moulder to SEL, December 1, 1967, 1–2, ASM Folder 11, Series I, SEL Papers, APS.

43. Moulder to SEL, December 1, 1967, 2.

44. Letter from Riley D. Housewright, Fort Detrick Technical Director, to SEL, March 28, 1968, ASM Folder 13, Series I, SEL Papers, APS. Housewright was the ASM president in 1966. (See Timeline of Officers on ASM website, https://lib.guides.umbc.edu/c.php?g=836720&p=6543752#s-lg-box-wrapper-24451675.)

45. Housewright to SEL, March 28, 1968.

46. "Report of the Council and CPC" *ASM News* 34, no. 3 (August 1968): 14.

47. Memo from Luria to CPC Members, Dr. E. M. Foster and Dr. J. Moulder, May 14, 1968, 1, ASM Folder 14, Series I, SEL Papers, APS.

48. Luria, "Microbiologist and His Times." See also the draft in SEL Papers, APS.

49. Luria, "Microbiologist and His Times," 401. Schalk has demonstrated how Camus was one of the influential French intellectuals who opposed the Algerian war, and how his position was admired by many of the American intellectuals who spoke out against Vietnam. Schalk, *War and the Ivory Tower*, 61–71.

50. Luria, "Microbiologist and His Times," 402.

51. Luria, "Microbiologist and His Times," 402–403.

52. Text of "Microbiologist and His Times" included in memo from SEL to CPC members, Dr. E. M. Foster, and Dr. J. Moulder, May 14, 1968, enclosure #1, ASM Folder 14, Series I, SEL Papers, APS.

53. Luria, "Microbiologist and His Times," 403.

54. *New York Times*, "Tie to Army Ended by Biology Society." The first quote was given with quotation marks, but the second one was not, the way it appears in the ASM press release. Press release, May 8, 1968, ASM, included in memo from SEL to CPC Members, Dr. E. M. Foster, and Dr. J. Moulder.

55. "Report of the Annual Business Meeting, 8 May 1968," *ASM News* 34, no. 3 (August 1968): 17.

56. Nelson, "Micro-Revolt of the Microbiologists," 862.

57. Nelson, "Micro-Revolt," 862.

58. Memo from SEL to CPC Members, Dr. E. M. Foster, and Dr. J. Moulder, 2. See also a handwritten letter to SEL from a young graduate student in support of his position, dated May 9, 1968, ASM Folder 14, Series I, SEL Papers, APS.

59. Press release, May 9, 1968, ASM Folder 14, Series I, SEL Papers, APS.

60. Memo from SEL to CPC Members, Dr. E. M. Foster, and Dr. J. Moulder, May 14, 1968, 3, ASM Folder 14, Series I, SEL Papers, APS.

61. Covert, *Cutting Edge*, 51–53; Miller, Engelberg, and Broad, *Germs*, 63.

62. On the Cancer Act, see chapter 6. On the conversion of Fort Detrick to a cancer center, see Rettig, *Cancer Crusade*, 278.

63. Nelson, "Micro-Revolt," 862.

64. Nelson, "Micro-Revolt," 862.

65. Leslie, *Cold War and American Science*, 14–15.

66. Weinberg, "Federal Laboratories and Science Education," 27–30, at 30n4. Also cited in Nelkin, *University and Military Research*, 24.

67. Murray Eden, "Historical Introduction," to Allen, *March 4*, vii–xxi; Nelkin, *University and Military Research*, 7; Moore, *Disrupting Science*, 137–146.

68. Eden, "Historical Introduction," in Allen, *March 4*, xiii. MIT professors were a very active group in the 1960s. As Richard Todd commented, "Professor-Politicians . . . are no rarity at MIT." Maury Fox recalls that faculty meetings were held in Kresge Auditorium, and that close to the entire faculty of 800 attended for the freewheeling discussions. Todd, "The 'Ins' and 'Outs' at MIT"; author interview with Maurice Fox, July 24, 2001.

69. "Faculty Sponsors of Activities on March 4," in Allen, *March 4*, xxiv–xxv.

70. Faculty Statement, January 27, Union of Concerned Scientists, Science Action Coordinating Committee Papers, AC 349, MIT Archive; Nelson, "Scientists Plan Research Strike."

71. Eden, "Historical Introduction," in Allen, *March 4*, xv; Nelkin, *Moral Politics*, 58–59.

72. Eden, "Historical Introduction," in Allen, *March 4*, xvii; Kendall, *Distant Light*, 2–3; Moore, *Disrupting Science*, 140–145.

73. A draft of the March 4 program dated February 7, 1969, listed all of the members by first name, including "SALVA" (Luria), "NOAM" (Chomsky), and "VICKI" (Weisskopf). Science Action Coordinating Committee Papers, AC 349, MIT Archive.

74. Magasanik, Ross, and Weisskopf, "No Research Strike at MIT."

75. SACC March 4 Student Statement, n.d., 3, Science Action Coordinating Committee Papers, AC 349, MIT Archive; Nelson, "MIT's March 4."

76. Nelson, "MIT's March 4," 1175.

77. Leslie, *Cold War and the University*, 234.

78. Program for March 4 and March 8, Union of Concerned Scientists Memo, 21 February 1969, 2, Science Action Coordinating Committee Papers, AC 349, MIT Archives.

79. Eugene Rabinowitch began his talk with the comment about how "our chairman . . . has wisely said that he is not going to participate in the discussion." Rabinowitch, "A Historical Perspective," in Allen, *March 4*, 65. The other participants were one student leader, two historians, and an economist.

80. Salloch, "Cambridge: March 4," 34. This is the only mention of the Saturday discussions in any of the journalistic accounts of the March 4 event. Neither the *New York Times* nor the *Boston Globe* covered the Saturday discussions.

81. Vietnam Moratorium Folder, Mendelsohn Papers.

82. Poster advertising an open meeting at Sanders Theater, Tuesday, October 14, 1969. Vietnam Moratorium Folder, Mendelsohn Papers.

83. Transcript of Luria's Nobel press conference, October 16, 1969, in S. E. Luria Biographical Folder, M.I.T. Museum Collection, Cambridge, MA. Hereafter MIT Museum Collection.

84. Author interview with Zella Hurwitz Luria, January 15, 2000.

85. Luria, *Slot Machine*, 192.

86. Nobel Prize Folders, Series III, SEL Papers, APS.

87. *Boston Globe*, October 17, 1969.

88. Text of both telegrams, dated October 17, 1969, in Richard M. Nixon folder, Series I, SEL Papers, APS.

89. Lyons, "Second H. E. W. Blacklist Includes Nobel Laureate"; Black, "Luria Never Consulted by U.S."; and Edstrom, "Nobelist Is on HEW Blacklist." See also *Life*, "Blacklisting Lingers On," 46; and *Newsweek*, "Blacklist's Backlash," 71.

90. Nelson, "Scientists Increasingly Protest"; and Nelson, "HEW Security Checks."

91. "Statement by S. E. Luria," October 20, 1969, News Releases Folder, MIT Museum Collection.

92. See, for example, "No Blacklists Needed," an editorial in the *Chicago Daily News*, October 12, 1969. The editors called the blacklist "furtive, irrelevant, mischievous, and un-American."

93. Smith, "No Tears for Luria."

94. Nelson, "HEW: Blacklists Scrapped."

95. Frank Annunzio, "Italian-Born Salvador E. Luria Wins Nobel Prize," Congressional Record—House, October 22, 1969, 31069, https://www.govinfo.gov/content/pkg/GPO-CRECB-1969-pt23/pdf/GPO-CRECB-1969-pt23-2-2.pdf. Letter from Frank Annunzio, October 26, 1969, Nobel Prize Folder #6, 22–27 October, Series I, SEL Papers, APS.

96. Annunzio, "Italian-Born Salvador E. Luria Wins Nobel Prize," 31069.

97. Luria, *Slot Machine*, 197.

98. Lanza, "Salvador Luria."

99. Women Strike for Peace invitation to a dinner honoring Luria, April 24, 1970, Women Strike for Peace Folder, Series IIa, SEL Papers, APS.

100. "Appeal to Members of the American Association for the Advancement of Science *For Science in the Service of Life*," n.d., Vietnam Folder, Mendelsohn Papers, Harvard Archives. The other members of the committee were John Edsal (Harvard), E. E. Pfeiffer (University of Montana), Arthur Galston (Yale), Arthur Westing (director of AAAS Herbicide Assessment), and Richard Lewontin (University of Chicago).

101. *New York Times*, "An Appeal: We Must Tell the President."

102. Lyons, "Scientists Assail Bombing Policy."

103. Bruce V. Lewenstein, "Shifting Science from People to Programs," in Kohlstedt, Sokal, and Lewenstein, *Establishment of Science in America*, 138–139; Moore, *Disrupting Science*, 165–169.

104. Gillette, "AAAS Meeting: Policy Change," 164.

105. Lyons, "Scientists Assail," 13.

106. Mendelsohn recalls that he was able to slip the motion in after the council had approved a motion condemning the U.S. Army for using live horses to do research on the damage done by land mines. If they had spoken out to protect the horses, then of course they would support a resolution protecting people and the environment. Comments made by Everett Mendelsohn in a panel discussion at the History of Science Society meeting, Denver, CO, November 9, 2001.

107. Gillette, "AAAS Council Meeting," 259.

108. Gillette, "AAAS Council Meeting," 259. Meselson was a major figure in the scientific outcry against chemical and biological weapons. See, for example, Miller, Engelberg, and Broad, *Germs*, chaps. 2 and 3.

109. Luria, "Bought a TV." The AP picked the letter up, and it was a news item in newspapers across New England and in the *New York Times*.

110. Letter from S.E. Luria to Brown University Board of Trustees, May 29, 1973, Brown University Folder, Series I, SEL Papers, APS.

111. Pave, "Peace Groups Protest Kissinger Prize."

112. Memorandum to FBI Director from SAC Boston, January 29, 1960, Letterhead Memorandum on Salvador E. Luria for the State Department, and Legats in Paris and Rome, February 9, 1967, 8–9. Despite the fact that no Communist informants in the area knew him, he was still classified as a Security Matter–C. FBI-SEL.

113. Letter to J. Edgar Hoover from JM, May 15, 1969, SEL-FBI.

114. Letter to President Nixon from a Boston area resident, December 30, 1972, FBI-SEL.

115. For a review of the antiscience movement of the 1960s, see Mendelsohn, "Politics of Pessimism."

BIOLOGY FOR AMERICAN SOCIETY

1. See Rettig, *Cancer Crusade*; Groopman, "Thirty Years' War"; Scheffler, *Contagious Cause,* chapter 7.

2. On the response to the budget cuts, see letter from Luria to Senator Edward Kennedy, September 24, 1969, Kennedy Folder, Series I, SEL Papers, APS. On Lasker, see Rettig, *Cancer Crusade*, chap. 2; and Groopman, "Thirty Years' War," 52.

3. National Program for the Conquest of Cancer: Report of the National Panel of Consultants on the Conquest of Cancer authorized by S. Res. 376. Prepared for the

Committee on Labor and Public Welfare, U.S. Senate (Washington, DC: U.S. Government Printing Office, 1971).

4. Holleb, "Comments from the Former Chief Medical Officer." See also Rettig, *Cancer Crusade*, chap. 5; Bazell, "Cancer Research Proposals"; Scheffler, *Contagious Cause*, 151–160.

5. See Rettig, *Cancer Crusade*, chaps. 4–8.

6. Rettig, *Cancer Crusade*, 293. The most prominent defector from the advisory panel was Joshua Lederberg, who openly criticized the proposed legislation in March. Brazell, "Lederberg Opposes Cancer Authority."

7. Rettig, *Cancer Crusade*, 143–151, 290. See also letters to the editor of the *New York Times* and the *Washington Post* from a group of scientists, including Arthur Kornberg, Marshall Nirenberg, and Severo Ochoa, criticizing Senator Javits's pride in the bill, August 9, 1971 (*NYT*) and September 5, 1971 (*WP*).

8. The lone dissenter, Senator Gaylord Nelson of Wisconsin, voted against the bill because of his own reservations as well as the urging of his constituents at the University of Wisconsin Medical School. Rettig, *Cancer Crusade*, 196.

9. Rettig, *Cancer Crusade*, 199–200.

10. Rettig, *Cancer Crusade*, 203. See also "Statement of Dr. Salvador E. Luria, Professor of Biology and Institute Professor, Massachusetts Institute of Technology."

11. Rettig, *Cancer Crusade*, 238, 258. Rettig claims that Rogers knew Luria would be sympathetic because he had publicly criticized the Senate bill in the *New York Times* and the *Washington Post*, but I was unable to find any editorials or letters to the editor with Luria's name on them, or any quotes from him in articles about the bill.

12. "Statement of Dr. Salvador E. Luria," 485. He did, however, take one of his fellow scientists to task for attending the signing of the National Cancer Act, chastising Ernest Borek in a letter thanking him for a copy of Borek's book. He wrote, "It appears that you went to Washington for the signing of the Cancer Bill. How could you stand it in the same room as that bastard? Cordially yours, S. E. Luria." Luria to Ernest Borek, February 22, 1974, Ernest Borek Folder, Series I, SEL Papers, APS.

13. "Statement of Dr. Salvador E. Luria," 486.

14. "Statement of Dr. Salvador E. Luria," 486.

15. "Statement of Dr. Salvador E. Luria," 486.

16. "Statement of Dr. Salvador E. Luria," 487.

17. "Statement of Dr. Salvador E. Luria," 488.

18. Hastings's comments recorded in "Statement of Dr. Salvador E. Luria," 489.

19. Letters to Senator Edward Kennedy and Senator Jacob Javits, November 22, 1971, Kennedy and Javits Folders, Series I, SEL Papers, APS.

20. Rettig, *Cancer Crusade*, 279; "The National Cancer Act of 1971," text of the law reprinted in "25th Anniversary of the Signing of the National Cancer Act," special section of *Cancer* 78, no. 12 (December 15, 1996): 2611.

21. Rettig, *Cancer Crusade*, 276.

22. "National Cancer Act of 1971," 2612.

23. "National Cancer Act of 1971," 2613.

24. See Studer and Chubin, *Cancer Mission*, for a sociological analysis of Baltimore's work that locates it somewhat in the context of cancer research. See also Crotty, *Ahead of the Curve*, 75–84. Howard Temin made a parallel but independent discovery at the same time.

25. Luria, "A Program in Cancer Cell Biology" draft, August 27, 1971, 1, and "Research Plan" in Grant Application, 9, AC 8, Box 53, Cancer Research Folder, MIT Archives.

26. Luria, "Program in Cancer Cell Biology," 2.

27. Luria, "Program in Cancer Cell Biology," 2.

28. Luria, "Program in Cancer Cell Biology," 3–4.

29. Luria, "Research Plan," 12. The MIT press office, in its release on "Background on Cancer Research at MIT," pointed out that research on cancer treatment and basic biology had been going on "for decades." Press release dated December 1, 1972, Cancer Research Folder, MIT Archives.

30. Luria, "Research Plan," 13–17.

31. Luria, "Research Plan," 13.

32. Crotty, *Ahead of the Curve*, 85.

33. Luria, "Program in Cancer Cell Biology," 11–12.

34. "Major Cancer Laboratory Announced for MIT," press release, December 5, 1972, AC 8, Cancer Research Folder, MIT Archives.

35. Luria, *Slot Machine*, 136, 138.

36. Letter from Luria to Delbrück, June 7, 1973, Delbrück Papers.

37. "An Emotional Attachment," in Inglis, Sambrook, and Witkowski, eds., *Inspiring Science*, 227–229.

38. Watson, "Director's Report"; Sharp, "Jim as a Mentor."

39. Brownlee, "Biography of Nancy Hopkins," 12790; Phillip Sharp Nobel Prize Biographical, https://www.nobelprize.org/prizes/medicine/1993/sharp/biographical/.

40. Sharp Nobel Prize Biographical; Sharp, "Life Sciences at MIT."

41. On Sharp's Nobel Prize–winning research, see Morange, *History of Molecular Biology*, 204–205; and Travis, "In Stockholm, a Clean Sweep." Sharp succeeded Luria as the director of the CCR when Luria retired in 1985, and was named the first Salvador E. Luria Professor of Biology in 1992.

42. Smigel, "1971–1991"; Luria, *Slot Machine*, 137. On Weinberg's research, see Morange, *History of Molecular Biology*, 223–227; Marx, "Cancer Cell Genes"; and Marx, "Oncogenes Reach a Milestone."

43. Brownlee, "Biography of Nancy Hopkins," 12789–12790.

44. Luria, *Slot Machine*, 137.

45. Groopman, "Thirty Years' War"; Scheffler, *Contagious Cause*.

46. Luria, "Modern Biology." Luria made that comment several years later in Luria, "Ethical and Institutional Aspects," 58.

47. Luria, "Modern Biology," 408–409.

48. Luria, "Modern Biology," 409.

49. Luria, "Molecular Biology: Past, Present, and Future," 1293, 1296.

50. Letter from Theodosius Dobzhansky to SEL, March 7, 1969, Dobzhansky Folder, Series I, SEL Papers, APS.

51. Luria, *Life: The Unfinished Experiment*, 3. He was committed to making his book accessible, in contrast to Jacques Monod's *Chance and Necessity: An Essay on the Natural Philosophy of Modern Biology*. This book is quite abstract and theoretical but was a best-seller in France and the United States.

52. Letters from Dobzhansky, June 4, September 11, and September 26, 1973, Dobzhansky Folder, Series I, SEL Papers, APS.

53. Letter to Scribner's, October 24, 1972, Scribner's Folder, Series I, SEL Papers, APS.

54. Luria, *Life*, 74.

55. Luria, *Life*, 3.

56. Luria, *Life*, 150.

57. Luria, *Life*, 4–5.

58. Luria, *Life*, 112, 126–127.

59. Author interview with Alice Huang, August 21, 2001; Luria, "Domestic Cooperation," n.d., Series III, SEL Papers, APS.

60. Luria, *Life*, 120.

61. Luria, *Life*, 122–125.

62. Luria, *Life*, 125–126.

63. Luria, *Life*, 85.

64. Luria, *Life*, 134.

65. E. O. Wilson's *Sociobiology* wasn't published until 1975, but there were other contemporary popular works on aggression and other human traits that gave biological or evolutionary explanations that can be characterized as sociobiological. The term itself had been coined in the 1940s by Ashley Montagu. See Sperling, "Ashley Montagu (1905–1999)."

66. Luria, *Life*, 134.

67. Luria, *Life*, 148.

68. Glass, "Review of *Life*"; Morrison, "Review of *Life*."

69. Edelson wrote a joint review of Luria's book and a book by Jonas Salk called *The Survival of the Wisest*, and he did not like Salk's vagueness and forays into "metabiology." Edelson, "Review of *Life* and *Survival of the Wisest*."

70. Letter from Dobzhansky to SEL, June 4, 1973, Dobzhansky folder; and letters from René Dubos to SEL, October 17, 1972, response from SEL, October 24, 1972, and answer October 31, 1972, Dubos Folder, both in Series I, SEL Papers, APS.

71. Kenny, "MIT's Luria"; McKeon, "Gene Drives a Hard Biological Bargain"; Harris, "Intricate Beauty of Our Molecules"; Klivington, "Biology for the Layman"; Stent, "Genes to Cells"; Kirsch, "World Beyond the Lab."

72. McKeon, "Gene Drives," 21.

73. Stent, "Genes to Cells," 28.

74. Klivington, "Biology," 18.

75. Klivington, "Biology," 19.

76. Kenny, "MIT's Luria," 41; letter from SEL to Dubos, October 24, 1972. Ironically, Luria refers quite often to Biblical imagery, such as the Garden of Eden.

77. Wiseman, "World of Books."

78. Baker, "The Awards," 43. See also Olson, "National Book Award Show."

79. Luria, "Acceptance Speech for the National Book Awards in the Sciences," National Book Award Statement, SEL Papers, APS.

80. Cole, "Best of All 25 N.B.A.'s."

81. Luria, "Domestic Cooperation."

82. Luria, "Domestic Cooperation," 1; Luria, *Slot Machine*, 53.

83. Luria, "Domestic Cooperation," 2.

84. Luria, "Domestic Cooperation," 3.

85. Luria, "Domestic Cooperation," 4–7. The recipe involves scrambling eggs with milk and salt, adding dry vermouth, and then tossing them with cold cream cheese.

86. Luria, "Domestic Cooperation," 8.

87. Luria, "Domestic Cooperation," 9.

88. Luria, "What Can Biologists Solve?" This article was a version of a talk he had given at the University of Rochester's Wilson Day celebration in the fall of 1973. See Luria, "Alternative Tasks," manuscript in "Alternative Tasks for Biologists" folder, Series III, SEL Papers, APS.

89. Luria, "What Can Biologists Solve?"

90. "Biologizing" is Luria's term.

91. Luria, "What Can Biologists Solve?" On the history of the debate over race and IQ, see Gould, *Mismeasure of Man*; Lewontin, Rose, and Kamin, *Not in Our Genes*, chap. 5; Fancher, *Intelligence Men*; and the documents collected in the "Root and Branch" section of Jacoby and Glauberman, *Bell Curve Debate*.

92. Luria, "What Can Biologists Solve?"

93. Luria, "What Can Biologists Solve?"

94. Bosworth, "How Useful Is IQ?" Bosworth's educational philosophy that he put into place at Saint Ann's explains his reasons for criticizing Luria's article. According to the school's website, Bosworth was committed to the use of "psychometrics, beginning with an eclectic admissions test and continuing with annual verbal and quantitative objective testing" in order to "prevent the tyranny of subjective judgment by authority figures." Students at Saint Ann's are not given grades—instead, teachers write detailed reports on each child that discuss their academic, social, and intellectual development. (Personal communication with Rebecca Perlin, Saint Ann's teacher from 1998 to 2001.) Therefore, it is not surprising that he would take issue with Luria's criticism of IQ tests, although Bosworth seems to feel that Luria's critique extended to any evaluation of intelligence, rather than the use of flawed tests to make scientific claims about social status. For more information about Saint Ann's, see https://saintannsny.org/about/history/.

95. Luria was not identified as an M.D. or a Nobel laureate in the magazine, and so Bosworth may not have known about Luria's stature in the scientific community.

96. Bosworth, "How Useful Is IQ?"

97. Luria, "Reply."

98. Montagu, *Man's Most Dangerous Myth*; Sperling, "Ashley Montagu (1905–1999)"; Graves, *Emperor's New Clothes*, 162–163.

99. Letter to the editor of the *New York Review of Books* from Ashley Montagu, April 19, 1974, New York Review of Books Correspondence Folder, Series I, SEL Papers, APS. For more on the history of how flawed and racist IQ tests are, see Gould, *Mismeasure of Man*, chap. 5.

100. Letter to R. B. Silvers from SEL, May 24, 1974, New York Review of Books Correspondence Folder, Series I, SEL Papers, APS.

101. Hall, "Italian-Americans Coming into Their Own," quote from Luria on 52. Luria's picture was part of a collage of famous Italian Americans on the cover of the *New York Times Magazine*. He was also featured on page 65 of the July 8, 1985, special issue of *Time* magazine on "Immigrants: The Changing Face of America."

102. Letter from SEL, February 21, 1986, Montedison Folder, Series I, SEL Papers, APS.

103. Luria, "Testimony provided for February 11 appearance before the House Subcommittee on Science, Research and Technology," January 28, 1976, United States House of Representatives Correspondence Folder, Series I, SEL Papers, APS; Luria, "Reflections on Democracy"; Luria, "Goals of Science"; Luria, "Single Artificer."

104. See Luria's publication list in his last Curriculum Vitae, Series IIa, SEL Papers, APS.

105. Author interview with Michael Weiss, October 1997.

106. MIT, the Equal Opportunity Committee, three folders, Series I, SEL Papers, APS.

107. Luria served as a consultant and was paid in stock. Repligen Folder in Series IIb, SEL Papers, APS.

108. Cook-Deegan, *Gene Wars*, 107–109.

109. Dulbecco, "Turning Point in Cancer Research."

110. Watson, "Human Genome Project," 46; Cook-Deegan, *Gene Wars*, "Part Two: Origins of the Genome Project," 79–116.

111. Koshland, "Sequences and Consequences."

112. Luria, "Human Genome Program."

113. Robert Cook-Deegan seems to see this as a veiled—and misguided—jibe at Watson. Cook-Deegan, *Gene Wars*, 171, 184.

114. Luria, "Human Genome Program."

115. Luria, "Human Genome Program."

116. Kevles, "Out of Eugenics." The main strand of opposition to the project at this point focused on increased competition for research funds, and whether or not it was

appropriate for biology to engage in a "Big Science" project that would interfere with other research initiatives. See Cook-Deegan, *Gene Wars*, 171–173.

117. Koshland, "Response."

118. Letter from David Baltimore to SEL, November 3, 1989, David Baltimore Folder, Series I, SEL Papers, APS.

119. See Crotty, *Ahead of the Curve*, 208–209; Kevles, "Out of Eugenics," 25; Watson, "Human Genome Project," 49.

120. Author interview with James D. Watson, January 17, 2002.

WINTER 1991, WINTER 2021

1. Cairns, Overbaugh, and Miller, "Origins of Mutants"; letters from Cairns to Luria, December 17 and 19, 1990, and letter from Luria to Cairns, December 19, 1990, John Cairns Folder, Series I, SEL Papers, APS. Evelyn Fox Keller discusses the implications of Cairns, Overbaugh, and Miller in "Between Language and Science."

2. Letter from Temporary Faculty Committee for Peace in the Persian Gulf to William Weld, December 27, 1990, Persian Gulf War Folder, Series IIa, SEL Papers, APS.

3. From the citation read by George H. W. Bush, September 16, 1991, quoted on National Medal of Science website, https://nationalmedals.org/laureate/salvador-e-luria/.

4. Memorandum from FBI, January 24, 1991, FBI-SEL.

5. Letter from Zella Luria to friends and colleagues, February 26, 1991, Memorial Service Folder, Series IIb, SEL Papers, APS.

6. Severo, "Salvador E. Luria Is Dead at 78."

7. Letters from David Baltimore, David Botstein, Stanley Hattman, and others, Condolence Letters Folders, Series IIb, SEL Papers, APS.

8. Condolence letter from Sue Neiman Offner, March 21, 1991, Condolence Letters Folders, Series IIb, SEL Papers, APS.

9. Letter from Gerald Holton to Zella Luria, February 11, 1991, Condolence Letters Folders, Series IIb, SEL Papers, APS.

10. Letter from Renato Dulbecco, February 8, 1991, Condolence Letters Folders, Series IIb, SEL Papers, APS.

11. Condolence letters from James Darnell and James Watson, February 6, 1991, Condolence Letters Folders, Series IIb, SEL Papers, APS; author interview with James D. Watson, January 17, 2002.

12. Watson, "Salvador E. Luria (1912–1991)." Reprinted in Watson, *A Passion for DNA*. I am grateful to Dr. Watson for sending me a reprint of this obituary.

13. Letter from Zella Luria to friends and colleagues, February 26, 1991, Memorial Service Folder, Series IIb, SEL Papers, APS.

14. Memorial Service Program, Memorial Service Folder, Series IIb, SEL Papers, APS.

15. "Salvador Edward Luria" resolution by the National Academy of Sciences, Condolence Letters Folder 3, Series IIb, SEL Papers, APS.

16. "Luria" statement by David Baltimore at Rockefeller University Commencement, June 5, 1991, Rockefeller University Correspondence Folder, Series I, SEL Papers, APS.

17. Handwritten note dated June 1, 1994, in National Medal of Science Folder, Series I, SEL Papers, APS.

18. *MIT Tech Talk*, "MIT Honors Luria, Magasanik, Sharp."

19. Koch Institute for Integrative Cancer Research at MIT, "Golden Anniversary for Luria's Gold Medal," October 20, 2019, https://ki.mit.edu/news/2019/golden-anniversary-lurias-gold-medal.

BIBLIOGRAPHY

ARCHIVAL COLLECTIONS

American Society for Microbiology Records and Salvador E. Luria interview with Leland McClung. American Society for Microbiology Archive, Baltimore.

Carnegie Institute of Washington Papers. Cold Spring Harbor Laboratory Archives, Cold Spring Harbor, NY.

John J. DeBoer Papers. University of Illinois Urbana-Champaign Archive.

Max Delbrück Papers. California Institute of Technology Archives, Pasadena.

Genetics Society of America Papers. American Philosophical Society Library, Philadelphia.

Irwin G. Gunsalus Papers. University of Illinois Urbana-Champaign Archive.

David D. Henry Presidential Papers. University of Illinois Urbana-Champaign Archive.

James R. Killian Papers. Massachusetts Institute of Technology, Institute Archives, Cambridge.

Salvador E. Luria FBI File, File Number 100–367221. Federal Bureau of Investigations, United States Department of Justice, Washington, DC.

Salvador E. Luria File. John Simon Guggenheim Foundation Archive, New York.

Salvador E. Luria INS File. Immigration and Naturalization Service, United States Department of Justice, Washington, DC.

Salvador E. Luria Papers. American Philosophical Society Library, Philadelphia.

Salvador E. Luria Publicity File. Massachusetts Institute of Technology, Institute Archives, MIT Museum, Cambridge.

Everett I. Mendelsohn Papers. Harvard University Archives, Cambridge, MA.

Lloyd Morey Presidential Papers. University of Illinois Urbana-Champaign Archive.

H. J. Muller Papers. Lilly Library Collection, Bloomington, IN.

Linus and Ava Helen Pauling Papers. Oregon State University, Corvallis.

Recombinant DNA Controversy History Collection. Massachusetts Institute of Technology, Institute Archives, Cambridge.

Rockefeller Foundation Records. Rockefeller Archive Center, Tarrytown, NY.

Tracy M. Sonneborn Papers. Lilly Library Collection, Bloomington, IN.

George G. Stoddard Presidential Papers. University of Illinois Urbana-Champaign Archive.

Julius A. Stratton Papers. Massachusetts Institute of Technology, Institute Archives, Cambridge.

James D. Watson Papers. Cold Spring Harbor Laboratory Archives, Cold Spring Harbor, NY.

Herman B. Wells Papers. Indiana University Archives, Bloomington.

PUBLISHED SOURCES

Note: Works by Luria used in this study are presented alphabetically. This does not comprise a complete list of Luria's published works, which is available in the American Philosophical Society Library collection.

Abir-Am, Pnina. "From Biochemistry to Molecular Biology: DNA and the Acculturated Journey of the Critic of Science Erwin Chargaff." *History and Philosophy of the Life Sciences* 2 (1980): 3–60.

Abir-Am, Pnina. "The Discourse of Physical Power and Biological Knowledge in the 1930s: A Reappraisal of the Rockefeller Foundation's 'Policy' in Molecular Biology." *Social Studies of Science* 12 (1982): 341–382.

Abir-Am, Pnina. "Essay Review: How Scientists View Their Heroes: Some Remarks on the Mechanism of Myth Construction." *Journal of the History of Biology* 15 (Summer 1982): 281–315.

Abir-Am, Pnina. "Themes, Genres, and Orders of Legitimation in the Consolidation of New Disciplines: Deconstructing the Historiography of Molecular Biology." *History of Science* 23 (1985): 73–117.

Abir-Am, Pnina. "Noblesse Oblige: Lives of Molecular Biologists." *Isis* 82 (1991): 326–343.

Abir-Am, Pnina. "The Politics of Macromolecules: Molecular Biologists, Biochemists, and Rhetoric." *Osiris* 7 (1992): 164–191.

Abir-Am, Pnina. "Entre mémoire collective et histoire en biologie moléculaire: les pre- miers rites commémoratifs pour les groupes fondateurs." In *La mise en mémoire de la science: pour une ethnographie historique des rites commémoratifs*, edited by Pnina G. Abir- Am, 25–74. Amsterdam: Overseas Publishers Association, 1998.

Abir-Am, Pnina. "The First American and French Commemorations in Molecular Biology: From Collective Memory to Comparative History." *Osiris* 14 (1999): 324–370.

Abir-Am, Pnina. "The Rockefeller Foundation and Refugee Biologists: European and American Careers of Leading RF Grantees from England, France, Germany, and Italy." In *The "Unacceptables": American Foundations and Refugee Scholars Between the Two World Wars and After*, edited by Giuliana Gemelli, 217–240. Brussels: Presses Interuniversita- ires Européennes, 2000.

Adamo, S. J. "From New York to Paris." *America* 119 (July 6, 1968): 20.

Adams, Mark B., ed. *The Evolution of Theodosius Dobzhansky: Essays on His Life and Thought in Russia and America*. Princeton, NJ: Princeton University Press, 1994.

"Against the Misuse of Science–An Appeal by M.I.T. Scientists." *Bulletin of the Atomic Scientists* 25 (February 1969): 8.

Allen, Garland E. *Life Science in the Twentieth Century*. New York: Cambridge Univer- sity Press, 1978.

Allen, Garland E. *Thomas Hunt Morgan: The Man and His Science*. Princeton, NJ: Prince- ton University Press, 1978.

Allen, Jonathan, ed. *March 4: Scientists, Students and Society*. Cambridge, MA: MIT Press, 1970.

Altman, Lawrence K. "Alfred D. Hershey, Nobel Laureate for DNA Work, Dies at 88." *New York Times*, May 24, 1997.

Amprino, Rodolfo. "Giuseppe Levi (1872–1965)." *Acta Anatomica* 66 (1967): 1–44.

Anderson, Thomas F. "Electron Microscopy of Phages." In *Phage and the Origins of Molecular Biology*, edited by John Cairns, Gunther Stent, and James D. Watson, 63–78. Cold Spring Harbor, NY: Cold Spring Harbor Laboratory Press, 1966, reprinted 1992.

Anderson, Thomas F. "Reflections on Phage Genetics." *Annual Review of Genetics* 15 (1981): 405–417.

Andrewes, C. H. "Discussion of Dr. Luria's Paper." *Cancer Research* 20 (June 1960): 689–694.

Angel, J. Lawrence. "Physical Anthropologists." *Science* 122 (July 1, 1955): 41.

Anissimov, Myriam. *Primo Levi: Tragedy of An Optimist*. Translated by Steve Cox. Wood- stock, NY: Overlook Press, 1999.

Appel, Toby. "Organizing Biology: The American Society of Naturalists and Its 'Affiliated Societies,' 1883–1923." In *The American Development of Biology*, edited by Ronald Rainger, Keith R. Benson, and Jane Maienschein, 87–120. Philadelphia: University of Pennsylvania Press, 1988.

Appel, Toby. *Shaping Biology: The National Science Foundation and American Biological Research, 1945–1975*. Baltimore: Johns Hopkins University Press, 2000.

Armor, David J., Joseph B. Giacquinta, R. Gordon McIntosh, and Diana E. H. Russell. "Professors' Attitudes Toward the Vietnam War." *Public Opinion Quarterly* 31 Summer 1967, 159–175.

Backscheider, Paula R. *Reflections on Biography*. New York: Oxford University Press, 1999.

Badash, Lawrence. "Science and McCarthyism." *Minerva* 38 (Spring 2001): 53–80.

Baker, John F. "The Awards: High Drama and Low Comedy." *Publishers Weekly*, May 13, 1974.

Baltimore, David, ed. *Nobel Lectures in Molecular Biology 1933–1975*. New York: Elsevier, 1977.

Baltimore, David. "Renato Dulbecco: A Biographical Memoir." *Biographical Memoirs of the National Academy of Sciences*. Washington, DC: National Academy of Sciences, 2014.

Barksy, Robert F. *Noam Chomsky: A Life of Dissent*. Toronto: ECW Press, 1997.

Bassani, Giorgio. *The Garden of the Finzi-Continis*. Translated by Isabel Quigly. New York: Atheneum, 1965.

Bazell, Robert J. "Cancer Research Proposals: New Money, Old Conflicts." *Science* 171 (March 5, 1971): 877–879.

Bazell, Robert J. "Lederberg Opposes Cancer Authority." *Science* 171 (March 26, 1971): 1220.

Bedarida, Guido. *Ebrei d'Italia*. Livorno: Societa Editrice Tirrena, 1950.

Berenson, Rose S. "Answer to Dr. Luria: Babi Yar and Warsaw Not Vietnam." *Jewish Advocate*, March 21, 1968.

Berry, R. Stephen, Mitio Inokuti, and A. R. P. Rau. "Ugo Fano, 1912–2001: A Biographical Memoir." *Biographical Memoirs of the National Academy of Sciences*. Washington, DC: National Academy of Sciences, 2009.

Bertani, Giuseppe. "Salvador Edward Luria (1912–1991)." *Genetics* 131 (1992): 1–4.

Bertani, Giuseppe. "Lysogeny at Mid-Twentieth Century: P1, P2, and Other Experimental Systems." *Journal of Bacteriology* 186 (2004): 595–600.

Bilensky, S. M., T. D. Bloskhintseva, I. G. Pokrovskaya, and M. G. Sapozhnikov, eds. *B. Pontevorvo Selected Scientific Works, Recollections on B. Pontecorvo*. Bologna: Società Italiana Fisica, 1997.

Black, Harbert. "Luria Never Consulted by U.S." *Boston Globe,* October 20, 1969.

Blatt, Joel. "The Battle of Turin, 1933–1936: Carlo Rosselli, Guistizia e Libertà, OVRA and the Origins of Mussolini's anti-Semitic Campaign." *Journal of Modern Italian Studies* 1 (1995): 22–57.

Bobbio, Norberto. *Trent'anni di storia della cultura a Torino, 1920–1950.* Turin: Cassa di Risparmio di Torino, 1977.

Bonner, James. "Editorial: Thoughts about Biology." *AIBS Bulletin* 10 (October 1960): 17.

Bonner, James. "The Future Welfare of Botany." *AIBS Bulletin* 13 (February 1963): 20–21.

Boston Globe. "Faculty for the Resistance." October 15, 1967.

Boston Globe. "G. I. Shamed by Suggestion We Quit Viet Nam." January 10, 1965.

Boston Globe. "He Promised It Twice and Still No Peace." December 28, 1972.

Boston Globe. "An Open Letter to President Johnson." December 18, 1964.

Boston Globe. "Scientists and Engineers for McCarthy." December 15, 1967.

Boston Globe. "What People Talk About: 'Negotiate All International Conflicts.'" February 27, 1965.

Bosworth, Stanley. "How Useful Is IQ?" *New York Review of Books.* May 2, 1974.

Boyer, Paul. *By the Bomb's Early Light: American Thought and Culture at the Dawn of the Atomic Age.* New York: Pantheon Books, 1985.

Braun, Werner. "Bacterial Dissociation—A Critical Review of a Phenomenon of Bacterial Variation." *Bacteriological Reviews* 11 (1947): 75–114.

Braun, Werner. "Studies on Bacterial Variation and Its Relation to Some General Biological Problems." *American Naturalist* 81 (1947): 262–275.

Brick, Howard. *Age of Contradiction: American Thought and Culture in the 1960s.* New York: Twayne, 1998.

Brock, Thomas D. *The Emergence of Bacterial Genetics.* Cold Spring Harbor, NY: Cold Spring Harbor Laboratory Press, 1990.

Brownlee, Christen. "Biography of Nancy Hopkins." *Proceedings of the National Academy of Sciences* 101 (2004): 12789–12791.

Buckley, William F. "Intellectuals' Protests of Vietnam War Analyzed." *Los Angeles Times*, September 2, 1966.

Buckser, Stanley. "Letter to the Editor." *Science* 158 (December 14, 1967): 1393.

Buzzanco, Robert. *Vietnam and the Transformation of American Life.* London: Blackwell Problems in American History Series, 1999.

Byrne, Kevin B., ed. *Responsible Science: The Impact of Technology on Society*. San Francisco: Harper and Row, 1986.

Cain, Joe. "The Columbia Biological Series, 1894–1974: A Bibliographic Note." *Archives of Natural History* 28 (2001): 353–366.

Cain, Joe. "Co-opting Colleagues: Appropriating Dobzhansky's 1936 Lectures at Columbia." *Journal of the History of Biology* 35 (2002): 207–219.

Cairns, John. *Matters of Life and Death: Perspectives on Public Health, Molecular Biology, Cancer, and the Prospects for the Human Race*. Princeton, NJ: Princeton University Press, 1997.

Cairns, John, Julie Overbaugh, and Stephan Miller. "The Origins of Mutants." *Nature* 335 (September 8, 1988): 142–145.

Cairns, John, Gunther Stent, and James D. Watson, eds. *Phage and the Origins of Molecular Biology*. Cold Spring Harbor, NY: Cold Spring Harbor Laboratory Press, 1966, revised and expanded edition, 1992.

"The Cambridge Experimentation Review Board." *Bulletin of the Atomic Scientists* 33 (May 1977): 22–26.

Carlson, Elof Axel. *Genes, Radiation and Society: The Life and Work of H. J. Muller*. Ithaca, NY: Cornell University Press, 1981.

Caute, David. *The Great Fear: The Anti-Communist Purge under Truman and Eisenhower*. New York: Simon & Schuster, 1978.

Chatfield, Charles. "The Antiwar Movement and America." In DeBenedetti, *An American Ordeal: The Antiwar Movement of the Vietnam Era*, with Charles Chatfield, 387–408. Syracuse, NY: Syracuse University Press, 1990.

Chicago Daily News. "No Blacklists Needed." October 21, 1969.

Chomsky, Noam. *American Power and the New Mandarins*. New York: Pantheon, 1967.

Chomsky, Noam. *Keeping the Rabble in Line: Interviews with David Barsamian*. Edinburgh: AK Press, 1994.

Clarke, Adele, and Joan Fujimura. *The Right Tools for the Job: At Work in Twentieth Century Life Sciences*. Princeton, NJ: Princeton University Press, 1992.

Cole, William. "The Best of All 25 N.B.A.'s." *New York Times Book Review*, May 5, 1974.

Collard, Patrick. *The Development of Microbiology*. New York: Cambridge University Press, 1976.

Comfort, Nathaniel. *The Tangled Field: Barbara McClintock's Search for the Patterns of Genetic Control*. Cambridge, MA: Harvard University Press, 2001.

Comfort, Nathaniel, and Bentley Glass. *Building Arcadia: A History of Cold Spring Harbor Laboratory*. Unpublished manuscript.

Comfort, Nathaniel, and Wendy Goldstein. *Coming of Phage: Celebrating the Fiftieth Anniversary of the First Phage Course*. Cold Spring Harbor, NY: Cold Spring Harbor Laboratory Press, 1995.

Cook-Deegan, Robert. *The Gene Wars: Science, Politics and the Human Genome*. New York: W. W. Norton, 1994.

Covert, Norman M. *Cutting Edge: A History of Fort Detrick, Maryland, 1943–1993*. Fort Detrick: U.S. Army Public Affairs Office, 1993.

Creager, Angela N. H. "Adaptation or Selection? Old Issues and New Stakes in the Postwar Debates over Bacterial Drug Resistance." *Studies in the History and Philosophy of Biology and the Biomedical Sciences* 38 (2007): 159–190.

Creager, Angela N. H. *The Life of a Virus: Tobacco Mosaic Virus as an Experimental Model*. Chicago: University of Chicago Press, 2002.

Creager, Angela N. H. "Mapping Genes in Microorganisms." In *From Molecular Genetics to Genomics: The Mapping Cultures of Twentieth-Century Genomics*, edited by Jean-Paul Gaudillière and Hans-Jörg Rheinberger, 9–41. New York: Routledge, 2004.

Creager, Angela N. H. "Virus Research in the Postwar Realignment of Medicine and Biology." Presented at "Biology and Medicine—A Russian-American Overview." American Philosophical Society, Philadelphia, August 10–12, 1998.

Cronin, James W., ed. *Fermi Remembered*. Chicago: University of Chicago Press, 2004.

Crotty, Shane. *Ahead of the Curve: David Baltimore's Life in Science*. Berkeley: University of California Press, 2001.

Crow, James F. "H. J. Muller's Role in Evolutionary Biology." In *The Founders of Evolutionary Genetics*, edited by Sahotra Sarkar. Boston: Kluwer Academic, 1992.

Culliton, Barbara J. "Recombinant DNA: Cambridge City Council Votes Moratorium." *Science* 193 (July 23, 1976): 300–301.

Culver, John C., and John Hyde. *American Dreamer: The Life and Times of Henry A. Wallace*. New York: W. W. Norton, 2000.

Cunningham, Ann Marie. "Nobel Winner Solves Life." *New York Times Book Review*, April 8, 1984.

Daston, Lorraine. "The Moral Economy of Science." *Osiris* 10 (1995): 1–24.

Davies, T. H. "Soviet Atomic Espionage." *Bulletin of the Atomic Scientists* 7 (May 1951): 143–148.

Davis, Bernard D. *Storm over Biology: Essays on Science, Sentiment and Public Policy*. Buffalo, NY: Prometheus Books, 1986.

Davis, James Kirkpatrick. *Assault on the Left: The FBI and the Sixties Antiwar Movement*. Westport, CT: Prager, 1997.

DeBenedetti, Charles. "On the Significance of Citizen Peace Activism: America, 1961–1975." *Peace and Change* 9 (Summer 1983): 6–20.

DeBenedetti, Charles, with Charles Chatfield. *An American Ordeal: The Antiwar Movement of the Vietnam Era.* Syracuse, NY: Syracuse University Press, 1990.

De Felice, Renzo. *Storia degli ebrei italiani sotto il fascismo.* Turin: Giulio Einaudi, 1961.

deJong-Lambert, William, and Nikolai Krementsov. "On Labels and Issues: The Lysenko Controversy and the Cold War." *Journal of the History of Biology* 45 (Fall 2012): 373–388.

Delbrück, Max, "Introductory Remarks about the Program." *Cold Spring Harbor Symposia on Quantitative Biology: Viruses* 18: 1–2. Cold Spring Harbor, NY: Cold Spring Harbor Laboratory Press, 1953.

Delbrück, Max, ed. *Viruses 1950.* Pasadena: Division of Biology of the California Institute of Technology, 1950.

Delbrück, Max, and Mary Bruce Delbrück. "Bacterial Viruses and Sex." *Scientific American*, November 1948.

Delbrück, Max, and S. E. Luria. "Interference Between Bacterial Viruses: I. Interference Between Two Bacterial Viruses Acting Upon the Same Host, and the Mechanism of Virus Growth." *Archives of Biochemistry* 1 (1942): 111–141.

Demerec, Milislav. "Annual Report of the Biological Laboratory of the Long Island Biological Association." 1945.

Demerec, Milislav. "Annual Report of the Biological Laboratory of the Long Island Biological Association." 1948.

Demerec, Milislav. "Origin of Bacterial Resistance to Antibiotics." *Journal of Bacteriology* 56 (1948): 63–74.

Demerec, Milislav. "Production of Staphylococcus Strains Resistant to Various Concentrations of Penicillin." *Proceedings of the National Academy of Sciences of the United States of America* 31 (1945): 16–24.

Demerec, Milislav, and Ugo Fano. "Bacteriophage-Resistant Mutants in *Escherichia-coli.*" *Genetics* 30 (1945): 119–136.

Demerec, Milislav, and S. E. Luria. "The Gene." *Carnegie Institute of Washington Year Book* 44 (1944–1945): 115–121.

Dolan, Eleanor F., and Margaret P. Davis. "Antinepotism Rules in American Colleges and Universities: Their Effect on the Faculty Employment of Women." *Educational Record* 41 (October 1941): 285–295.

"Dr. F. Holweck." *Nature* 149, no. 3771 (1942): 163.

Duckworth, Donna H. "Who Discovered Bacteriophage?" *Bacteriological Reviews* 40 (1976): 793–802.

Dulbecco, Renato. *Scienza, Vita e Avventura*. Milan: Sperling & Kupfer Editori, 1989.

Dulbecco, Renato. "A Turning Point in Cancer Research: Sequencing the Human Genome." *Science* 231 (March 7, 1986): 1055–1066.

Dupree, A. Hunter. *Science in the Federal Government: A History of Policies and Activities*. Baltimore: Johns Hopkins University Press, 1957, 1986.

Eakin, Emily. "On the Lookout for Patriotic Incorrectness." *New York Times*, November 24, 2001.

Edel, Leon. *Writing Lives: Principia Biographica*. New York: W. W. Norton, 1984.

Edelson, Edward. "Review of *Life: The Unfinished Experiment* and *Survival of the Wisest*." *Smithsonian* 4 (August 1973): 84.

Edgcomb, Gabrielle. *From Swastika to Jim Crow: Refugee Scholars at Black Colleges*. Malabar, FL: Krieger, 1993.

Edstrom, Eve. "Nobelist Is on HEW Blacklist." *Washington Post*, October 21, 1969.

Egar, R. S. "The Americanization of Luria." *Science* 225 (July 6, 1984): 47.

Ehrenreich, Barbara. *Fear of Falling: The Inner Life of the Middle Class*. New York: Harper Perennial, 1990.

Engelhardt, Tom. *The End of Victory Culture: Cold War America and the Disillusioning of a Generation*. New York: Basic Books, 1995.

Epstein, Abraham. "Mishpachat Luria (Luria Family)." In *Kitvei R'Avraham Epstein (The Works of Rabbi Abraham Epstein)*. Jerusalem: Mosad HaRav Kook, 1901(?).

Erenberg, Lewis, and Susan Hirsch, eds. *The War in American Culture: Society and Consciousness During World War II*. Chicago: University of Chicago Press, 1996.

Exner, F. M., and S. E. Luria. "The Size of Streptococcus Bacteriophages as Determined by X-Ray Inactivation." *Science* 94 (1941): 394–395.

Ezrahi, Yaron. *The Descent of Icarus: Science and the Transformation of Contemporary Democracy*. Cambridge, MA: Harvard University Press, 1990.

Fancher, Raymond E. *The Intelligence Men: Makers of the IQ Controversy*. New York: W. W. Norton, 1985.

Farber, David. "The Counterculture and the Antiwar Movement." In *Give Peace a Chance: Exploring the Vietnam Antiwar Movement*, edited by Melvin Small and William D. Hoover, 7–21. Syracuse, NY: Syracuse University Press, 1992.

Farber, David. *The Age of Great Dreams: America in the 1960s*. New York: Hill and Wang, 1994.

Farber, David, ed. *The Sixties: From Memory to History*. Chapel Hill: University of North Carolina Press, 1994.

Feffer, Stuart M. "Atoms, Cancer and Politics: Supporting Atomic Science at the University of Chicago, 1944–1950." *Historical Studies in the Physical Sciences* 22 (1992): 233–261.

Fenner, Frank. "The Poxvirus." In *Portraits of Viruses: A History of Virology*, edited by F. Fenner and A. Gibbs. New York: Karger, 1988.

Fermi, Laura. *Atoms in the Family: My Life with Enrico Fermi*. Volume 9 in the *History of Modern Physics, 1800–1950*. Woodbury, NY: American Institute of Physics/ Tomash Publishers, reprinted 1987.

Fermi, Laura. *Illustrious Immigrants: The Intellectual Migration from Europe, 1930–1941*. Chicago: University of Chicago Press, 1968.

Feyerabend, Paul. "Democracy, Elitism, and Scientific Method." *Inquiry* 23 (1980): 3–18.

Finch, Eleanor F. "59th Annual Meeting of the American Society for International Law." *American Journal of International Law* 59 (July 1965): 599–604.

Fischer, Ernst Peter, and Carol Lipson. *Thinking About Science: Max Delbrück and the Origins of Molecular Biology*. New York: W. W. Norton, 1988.

Fleming, Donald. "Émigré Physicists and the Biological Revolution." In *The Intellectual Migration: Europe and America, 1930–1960*, edited by Donald Fleming and Bernard Bailyn, 152–185. Cambridge, MA: Harvard University Press, 1968.

Forgacs, David. *Italian Culture in the Industrial Era, 1880–1980: Cultural Industries, Politics, and the Public*. Manchester, UK: Manchester University Press, 1990.

Fosdick, Raymond B. *The Story of the Rockefeller Foundation*. New York: Harper and Brothers, 1952.

Fox, Richard Wightman, and T. J. Jackson Lears, eds. *The Culture of Consumption: Critical Essays in American History, 1880–1980*. New York: Pantheon, 1983.

Fox, Richard Wightman, and T. J. Jackson Lears, eds. *The Power of Culture: Critical Essays in American History*. Chicago: University of Chicago Press, 1993.

Fox, Robert, and Anna Guagnini, eds. *Education, Technology and Industrial Performance in Europe, 1850–1939*. New York: Cambridge University Press, 1993.

Franklin, Richard M. "Virology after 14 Years of Discovery." *Science* 158 (October 27, 1967): 484.

Fredrickson, Donald S. *The Recombinant DNA Controversy, a Memoir: Science, Politics and the Public Interest, 1974–1981*. Washington, DC: ASM Press, 2001.

Frum, David. *How We Got Here: The 70's: The Decade That Brought You Modern Life (For Better or Worse)*. Toronto: Random House Canada, 2000.

Fujimura, Joan H. "Crafting Science: Standardized Packages, Boundary Objects, and 'Translation.'" In *Science as Practice and Culture*, edited by Andrew Pickering, 168–211. Chicago: University of Chicago Press, 1992.

Fujimura, Joan H. *Crafting Science: A Sociohistory of the Quest for the Genetics of Cancer.* Cambridge, MA: Harvard University Press, 1996.

Galison, Peter. *Image and Logic: A Material Culture of Microphysics.* Chicago: University of Chicago Press, 1997.

Garcia, Guy. "Ten Routes to the American Dream." *Time*, July 8, 1985.

Gardner, David. *The California Oath Controversy.* Berkeley: University of California Press, 1967.

Gaudilliére, Jean-Paul. *Biologie moleculaire et biologists dans les anees soixante: La Naissance d'une discipline: La cas francais.* University of Paris doctoral thesis, 1991.

Gaudilliére, Jean-Paul. "Molecular Biologists, Biochemists, and Messenger RNA: The Birth of a Scientific Network." *Journal of the History of Biology* 29 (1996): 417–445.

Gaudilliére, Jean-Paul. "Molecular Biology in the French Tradition? Redefining Local Traditions and Disciplinary Patterns." *Journal of the History of Biology* 26 (1993): 473–498.

Gaudilliére, Jean-Paul. "The Molecularization of Cancer Etiology in the Postwar United States: Instruments, Politics and Management." In *Molecularizing Biology and Medicine: New Practices and Alliances, 1910s–1970s*, edited by Soraya de Chadarevian and Harmke Kamminga, 139–170. Amsterdam: Harwood Academic Publishers, 1998.

Gieryn, Thomas F. *Cultural Boundaries of Science: Credibility on the Line.* Chicago: University of Chicago Press, 1999.

Gillette, Robert. "AAAS Council Meeting: Vietnam Resolutions; Bylaws Voted." *Science* 179 (January 19, 1973): 258–262.

Gillette, Robert. "AAAS Meeting: Policy Change on Activists Brings Police." *Science* 179 (January 12, 1973): 162–164.

Ginzburg, Natalia. *Family Sayings.* Revised from the original translation by D. M. Low. Manchester, UK: Carcanet, 1984.

Gitlin, Todd. *The Sixties: Years of Hope, Days of Rage.* New York: Bantam Books, 1987, 1993.

Glass, Bentley. "Review of *Life: The Unfinished Experiment.*" *Quarterly Review of Biology* 49 (September 1974): 247.

Goertzel, Ted, and Ben Goertzel. *Linus Pauling: A Life in Science and Politics.* New York: Basic Books, 1995.

Goldstein, Toby. *Waking from the Dream: America in the Sixties*. New York: Julian Messner, 1988.

Goodell, Rae. "Public Involvement in the DNA Controversy: The Case of Cambridge, Massachusetts." *Science, Technology and Human Values* 4 (Spring 1979): 36–43.

Goodell, Rae. *The Visible Scientists*. Boston: Little, Brown, 1975.

Gordin, Michael D. "How Lysenkoism Became Pseudoscience: Dobzhansky to Velikovsky." *Journal of the History of Biology* 45 (Fall 2012): 443–468.

Gormley, Melinda. "Scientific Discrimination and the Activist Scientist: L. C. Dunn and the Professionalization of Genetics and Human Genetics in the United States." *Journal of the History of Biology* 42 (2009): 33–72.

Gould, Stephen Jay. "The Hardening of the Modern Synthesis." In *Dimensions of Darwinism: Themes and Counterthemes in Twentieth-Century Evolutionary Theory*, edited by Marjorie Grene, 71–93. New York: Cambridge University Press, 1983.

Gould, Stephen Jay. *The Mismeasure of Man*. New York: W. W. Norton, 1996.

Grafe, Alfred. *A History of Experimental Virology*. Translated by Elvira Reckendorf. New York: Springer-Verlag, 1991.

Graebner, William. *The Age of Doubt: American Thought and Culture in the 1940s*. Boston: Twayne, 1991.

Graham, George J., Jr. "The Necessity of the Tension." *Social Epistemology* 7 (1993): 25–34.

Grant, William C., Jr., and George B. Saul II. "A Curricular 'Straw Man' for Biologists." *AIBS Bulletin* 10 (February 1960): 15–18.

Graves, Joseph L., Jr. *The Emperor's New Clothes: Biological Theories of Race at the Millennium*. New Brunswick, NJ: Rutgers University Press, 2001.

Grilli, Marcel. "The Role of the Jews in Modern Italy." *Menorah Journal* 27 (1939): 260–80, and 28 (1940): 60–81, 172–197.

Grobman, Arnold B. "The BSCS: A Challenge to the Colleges." *AIBS Bulletin* 11 (December 1961): 17–20.

Grobstein, Clifford. *A Double Image of the Double Helix: The Recombinant DNA Debate*. San Francisco: W. H. Freeman, 1979.

Groopman, Jerome. "The Thirty Years' War." *New Yorker*, June 4, 2001.

Guston, David. "The Essential Tension in Science and Democracy" and "Resolving the Tension in Graham and Laird." *Social Epistemology* 7 (1993): 3–23, 47–60.

Hager, Thomas. *Force of Nature: The Life of Linus Pauling*. New York: Simon & Schuster, 1995.

Hall, Mitchell K. "CALCAV and Religious Opposition to the Vietnam War." In *Give Peace a Chance: Exploring the Vietnam Antiwar Movement*, edited by Melvin Small and William D. Hoover, 22–34. Syracuse, NY: Syracuse University Press, 1992.

Hall, Stephen S. "Italian-Americans Coming into Their Own." *New York Times Magazine*, May 15, 1983.

Hamby, Alonzo L. *Liberalism and Its Challengers: From F.D.R. to Bush*. 2nd ed. New York: Oxford University Press, 1992.

Hankins, Thomas L. "In Defence of Biography: The Use of Biography in the History of Science." *History of Science* 17 (1979): 1–16.

Hanna, Kathi E., ed. *Biomedical Politics*. Washington, DC: National Academy Press, 1991.

Harman, Oren Solomon. "C. D. Darlington and the British and American Reaction to Lysenko and the Soviet Conception of Science." *Journal of the History of Biology* 36 (2003): 309–352.

Harman, Oren Solomon. *The Man Who Invented the Chromosome: A Life of Cyril Darlington*. Cambridge, MA: Harvard University Press, 2004.

Harris, Harold H. "The Intricate Beauty of Our Molecules." *St. Louis Post-Dispatch*, November 4, 1973.

Hartmann, Susan. *The Home Front and Beyond: American Women in the 1940s*. Boston: Twayne, 1982.

Heineman, Kenneth J. *Campus Wars: The Peace Movement at American State Universities in the Vietnam Era*. New York: New York University Press, 1993.

Heineman, Kenneth J. "'Look Out Kid, You're Gonna Get Hit!': Kent State and the Vietnam Antiwar Movement." In *Give Peace a Chance: Exploring the Vietnam Antiwar Movement*, edited by Melvin Small and William D. Hoover, 201–222. Syracuse: Syracuse University Press, 1992.

Herring, George C. *America's Longest War: The United States and Vietnam, 1950–1975*. 3rd ed. New York: McGraw-Hill, 1996.

Hershey, Alfred D. "Mutation of Bacteriophage with Respect to Type of Plaque." *Genetics* 31 (1946): 620–640.

Hershey, Alfred D., and Martha Chase. "Independent Functions of Viral Protein and Nucleic Acid in Growth of Bacteriophage." *Journal of General Physiology* 36 (September 20, 1952): 39–56.

Holland, John. "Review of *Virology*." *ASM News* 33 (August 1967): 68–69.

Holleb, Arthur I. "Comments from the Former Chief Medical Officer of the American Cancer Society (1968–1988)." *Cancer* 78 (December 15, 1996): 2590–2591.

Hollinger, David A. *Science, Jews and Secular Culture: Studies in Mid-Twentieth Century American Intellectual History*. Princeton, NJ: Princeton University Press, 1996.

Holmes, Frederic L. *Reconceiving the Gene: Seymour Benzer's Adventures in Phage Genetics*, edited by William C. Summers. New Haven, CT: Yale University Press, 2006.

Howard, Ted, and Jeremy Rifkin. *Who Should Play God? The Artificial Creation of Life and What It Means for the Future of the Human Race*. New York: Dell, 1977.

Hughes, H. Stuart. *Prisoners of Hope: The Silver Age of Italian Jews, 1924–1974*. Cambridge, MA: Harvard University Press, 1983.

Hughes, Sally Smith. *The Virus: A History of the Concept*. New York: Science History Publications, 1977.

Indiana Alumni Magazine. "Lecturer Abroad." October 1947.

Inglis, John R., Joseph Sambrook, and Jan A. Witkowski, eds. *Inspiring Science: Jim Watson and the Age of DNA*. Cold Spring Harbor, NY: Cold Spring Harbor Laboratory Press, 2003.

Irwin, M. R. "Proceedings of Meetings." *Records of the Genetics Society of America* 19 (1950): 24.

Jackson, David A., and Stephen P. Stich, eds. *The Recombinant DNA Debate*. Englewood Cliffs, NJ: Prentice Hall, 1979.

Jacob, Francois. *Of Flies, Mice, and Men*. Translated by Giselle Weiss. Cambridge, MA: Harvard University Press, 1998.

Jacob, Francois. *The Logic of Life: A History of Heredity*. New York: Pantheon Books, 1973.

Jacob, Francois. *The Statue Within: An Autobiography*. Translated by Franklin Philip. New York: Basic Books, 1988. Reprinted by the Cold Spring Harbor Press, 1995.

Jacobs, Charlotte DeCroes. *Jonas Salk: A Life*. New York: Oxford University Press, 2015.

Jacobson, Harold Karan, and Eric Stein. *Diplomats, Scientists, and Politicians: The United States and Nuclear Test Ban Negotiations*. Ann Arbor: University of Michigan Press, 1966.

Jacoby, Russell, and Naomi Glauberman, eds. *The Bell Curve Debate: History, Documents, Opinions*. New York: Time Books, 1995.

Jeffreys-Jones, Rhodri. *Peace Now! American Society and the Ending of the Vietnam War*. New Haven, CT: Yale University Press, 1999.

Joravsky, David. *The Lysenko Affair*. Cambridge, MA: Harvard University Press, 1970.

Judson, Horace Freeland. *The Eighth Day of Creation: Makers of the Revolution in Biology*. New York: Simon & Schuster, 1979.

Kadushin, Charles. *The American Intellectual Elite*. Boston: Little, Brown, 1974.

Kaiser, David. "Nuclear Democracy: Political Engagement, Pedagogical Reform, and Particle Physics in Postwar America." *Isis* 93 (2002): 229–268.

Katz, Milton S. "Peace Liberals and Vietnam: SANE and the Politics of 'Responsible' Protest." *Peace and Change* 9 (Summer 1983): 21–39.

Kay, Lily E. "Conceptual Models and Analytical Tools: The Biology of Physicist Max Delbrück." *Journal of the History of Biology* 18, no. 2 (Summer 1985): 207–247.

Kay, Lily E. "Life as Technology: Representing, Intervening and Molecularizing." In *The Philosophy and History of Molecular Biology: New Perspectives*, Boston Studies in the Philosophy of Science, vol. 183, edited by Sahotra Sarkar, 87–100. Boston: Kluwer Academic Publishers, 1996.

Kay, Lily E. *The Molecular Vision of Life: Caltech, the Rockefeller Foundation and the Rise of the New Biology*. Oxford: Oxford University Press, 1993.

Kay, Lily E. "Quanta of Life: Atomic Physics and the Reincarnation of Phage." *History and Philosophy of the Life Sciences* 14 (1992): 3–21.

Kay, Lily E. *Who Wrote the Book of Life? A History of the Genetic Code*. Stanford, CA: Stanford University Press, 2000.

Kay, Lily E. "W. M. Stanley's Crystallization of the Tobacco Mosaic Virus." *Isis* 77 (1986): 450–472.

Keith-Lucas, Thea. "Why Radius? A Letter to Friends of the Technology & Culture Forum." *Radius* (blog), October 13, 2014. https://radius.mit.edu/blog/why-radius-letter-friends-technology-culture-forum.

Keller, Evelyn Fox. *A Feeling for the Organism: The Life and Work of Barbara McClintock*. San Francisco: W. H. Freeman, 1983.

Keller, Evelyn Fox. "Between Language and Science: The Question of Directed Mutation in Molecular Genetics." *Perspectives in Biology and Medicine* 35 (Winter 1992): 292–306.

Keller, Evelyn Fox. *The Century of the Gene*. Cambridge, MA: Harvard University Press, 2000.

Kendall, Henry W. *A Distant Light: Scientists and Public Policy*. New York: Springer-Verlag, 2000.

Kenny, Herbert A. "MIT's Luria Sizes Up Molecular Biology Maze." *Boston Globe*, May 8, 1974.

Kevles, Daniel J. "Out of Eugenics: The Historical Politics of the Human Genome." In *The Code of Codes: Scientific and Social Issues in the Human Genome Project*, edited by Daniel J. Kevles and Leroy Hood. Cambridge, MA: Harvard University Press, 1992.

Kevles, Daniel J. "Renato Dulbecco and the New Animal Virology: Medicine, Methods, and Molecules." *Journal of the History of Biology* 26 (Fall 1993): 409–442.

Kevles, Daniel J., and Gerald L. Geison. "The Experimental Life Sciences in the Twentieth Century." *Osiris* 10 (1995): 139–163.

Kirsch, Robert. "A World Beyond the Lab." *Los Angeles Times*, October 1, 1974.

Kleinman, Mark L. *A World of Hope, a World of Fear: Henry A. Wallace, Reinhold Niebuhr, and American Liberalism.* Columbus: Ohio State University Press, 2000.

Klivington, Kenneth. "Biology for the Layman." *Los Angeles Times*, December 17, 1973.

Kohler, Robert E. *Lords of the Fly: Drosophila Genetics and the Experimental Life.* Chicago: University of Chicago Press, 1994.

Kohler, Robert E. "The Management of Science: Warren Weaver and the Rockefeller Programme in Molecular Biology." *Minerva* 14 (1976): 279–306.

Kohler, Robert E. *Partners in Science: Foundations and Natural Scientists, 1900–1945.* Chicago: University of Chicago Press, 1991.

Kohlstedt, Sally Gregory, Michael M. Sokal, and Bruce V. Lewenstein. *The Establishment of Science in America: 150 Years of the American Association for the Advancement of Science.* New Brunswick, NJ: Rutgers University Press, 1999.

Kolata, Gina. *Clone: The Road to Dolly, and the Path Ahead.* New York: Morrow, 1998.

Kolata, Gina. "What Is Warm and Fuzzy Forever? With Cloning, Kitty." *New York Times*, February 15, 2002.

Kornberg, Arthur. *For the Love of Enzymes: The Odyssey of a Biochemist.* Cambridge, MA: Harvard University Press, 1989.

Koshland, Daniel K., Jr. "Response." *Science* 246 (November 17, 1989): 873.

Koshland, Daniel K., Jr. "Sequences and Consequences of the Human Genome." *Science* 246 (October 13, 1989): 189.

Krementsov, Nikolai. "A 'Second Front' in Soviet Genetics: The International Dimension of the Lysenko Controversy, 1944–47." *Journal of the History of Biology* 29 (1996): 229–250.

Krimsky, Sheldon. *Genetic Alchemy: The Social History of the Recombinant DNA Controversy.* Cambridge, MA: MIT Press, 1982.

Kruger, D. H., and H. R. Gelderblom. "Helmut Ruska and the Visualization of Viruses." *Lancet* 355 (2000): 1713–1717.

Kutler, Stanley I. *The American Inquisition: Justice and Injustice in the Cold War.* New York: Hill and Wang, 1982.

Kuznick, Peter J. *Beyond the Laboratory: Scientists as Political Activists in 1930s America*. Chicago: University of Chicago Press, 1987.

Kuznick, Peter J., and James Gilbert, eds. *Rethinking Cold War Culture*. Washington, DC: Smithsonian Institution Press, 2001.

Lachmanovich, Liraz. "The Holocaust of the French Jews—A Historical Review." Yad Vashem, the World Holocaust Remembrance Center. https://www.yadvashem.org /articles/general/historical-review.html.

Ladd, Everett Carl, Jr. "Professors and Political Petitions." *Science* 163 (March 28, 1969): 1425–1430.

Laetsch, M. "The Welfare of Botany Re-Examined." *AIB Bulletin* 13 (December 1963): 21–22.

LaFollette, Marcel C. *Making Science Our Own: Public Images of Science, 1910–1955*. Chicago: University of Chicago Press, 1990.

Laird, Frank. "Participating in the Tension." *Social Epistemology* 7 (1993): 35–46.

Lanouette, William, with Bela Silard. *Genius in the Shadows: A Biography of Leo Szilard, the Man Behind the Bomb*. New York: Scribner's, 1992.

Lanza, Robert. "Salvador Luria: The Man Behind the Curtain." *Perspectives in Science and Medicine* 37 (Spring 1994): 359–361.

Latour, Bruno. "Give Me a Laboratory and I Will Raise the World." In *Science Observed*, edited by Karin D. Knorr-Cetina and Michael Mulkay, 141–170. London: Sage, 1983.

Lear, John. *Recombinant DNA: The Untold Story*. New York: Crown, 1978.

Lederman, Muriel, and Sue A. Tolin. "OVATOOMB: Other Viruses and the Origins of Molecular Biology." *Journal of the History of Biology* 26 (1993): 239–254.

Lee, John M. "3 Americans Get Nobel Prize in Medicine." *New York Times*, October 17, 1969.

Lepore, Jill. "Historians Who Love Too Much: Reflections on Microhistory and Biography." *Journal of American History* 88 (June 2001): 129–144.

Leslie, Stuart W. *The Cold War and American Science: The Military-Industrial-Academic Complex at MIT and Stanford*. New York: Columbia University Press, 1993.

Levi, Primo. "The Jews of Turin." Preface to *Ebrei a Torino—Ricirche per il Centenario della sinagoga 1884–1984*. Turin: Archivo di arte e cultura piedmontesi, Umberto Allemandi, 1984.

Levi, Primo. *The Periodic Table*. Translated by Raymond Rosenthal. New York: Schoken Books, 1984.

Levi-Montalcini, Rita. *In Praise of Imperfection: My Life and Work*. Translated by Luigi Attardi. New York: Basic Books, 1988.

Levi-Montalcini, Rita. "From Turin to Stockholm via St. Louis and Rio de Janeiro." *Science* 287 (February 4, 2000): 809.

Levy, David. *The Debate over Vietnam*. Baltimore: Johns Hopkins University Press, 1991, 2nd ed. 1995.

Lewontin, Richard C. "The Cold War and the Transformation of the Academy." In Noam Chomsky et al. *The Cold War and the University: Toward an Intellectual History of the Postwar Years*, 1–34. New York: New Press, 1997.

Lewontin, Richard C., Steven Rose, and Leon J. Kamin. *Not in Our Genes: Biology, Ideology and Human Nature*. New York: Pantheon, 1984.

Life. "Blacklisting Lingers On." November 7, 1969.

Lindee, M. Susan. *Suffering Made Real: American Science and the Survivors at Hiroshima*. Chicago: University of Chicago Press, 1994.

Lindgren, Carl. "Letter to the Editor." *Scientific American*, March 1950.

Lindquist, Clarence B., and Edwin L. Miller. "Degrees Conferred in the Biological Sciences." *AIBS Bulletin* 12 (February 1962): 28–32.

Lissner, Will. "6,000 in Colleges Sign Bomb Appeal: Protest to Johnson Reported from 200 Faculties." *New York Times*, January 22, 1967.

Lora, Ronald, ed. *America in the 1960s: Cultural Authorities in Transition*. New York: Wiley, 1974.

Los Angeles Times. "The Nixon Administration Is at War with America." October 25, 1972.

Luria, S. E. "Arrest Isn't Everything." *New York Times Magazine*, April 13, 1975.

Luria, S. E. "Babi Yar, Warsaw and Vietnam." *Jewish Advocate*, March 14, 1968.

Luria, S. E. "Bacteriophage: An Essay on Virus Reproduction." *Science* 111 (May 12, 1950): 507–511.

Luria, S. E. "Bacteriophage: An Essay on Virus Reproduction." In *Viruses 1950*, edited by Max Delbrück, 7–16. Pasadena: Division of Biology of the California Institute of Technology, 1950.

Luria, S. E. "Bought a TV." *Boston Globe*, November 25, 1969.

Luria, S. E. "Directed Genetic Change: Perspectives from Molecular Genetics." In *The Control of Human Heredity and Evolution*, edited by T. M. Sonneborn, 1–19. New York: Macmillan, 1965.

Luria, S. E. "Discussion of Dr. Muller's Paper." In *The Control of Human Heredity and Evolution*, edited by T. M. Sonneborn, 124. New York: Macmillan, 1965.

Luria, S. E. "Early Days in the Molecular Biology of Bacteriophage." In *Proceedings of the Conference on the History of Biochemistry and Molecular Biology*, edited by John Edsall, 22–28. Sponsored by the American Academy of Arts and Sciences, Brookline, MA, May 21–23, 1970.

Luria, S. E. "Ethical and Institutional Aspects of Recombinant DNA Technology." In *Recombinant DNA Research and the Human Prospect*, edited by Earl D. Hanson. Washington, DC: American Chemical Society, 1983.

Luria, S. E. "Few Raised Voices." *Boston Globe*, April 26, 1986.

Luria, S. E. "The Frequency Distribution of Spontaneous Bacteriophage Mutants as Evidence for the Exponential Rate of Phage Reproduction." *Cold Spring Harbor Symposia on Quantitative Biology: Genes and Mutations* 16: 463–470. Cold Spring Harbor, NY: Cold Spring Harbor Press, 1951.

Luria, S. E. *General Virology*. New York: Wiley, 1953.

Luria, S. E. "The Goals of Science." *Bulletin of the Atomic Scientists* 33 (May 1977): 28–33.

Luria, S. E. "Host-Induced Modifications of Viruses." *Cold Spring Harbor Symposia on Quantitative Biology: Viruses* 18: 237–244. Cold Spring Harbor, NY: Cold Spring Harbor Press, 1953.

Luria, S. E. "Human Genome Program." *Science* 246 (November 17, 1989): 873.

Luria, S. E. "A Latter-Day Rationalist's Lament." In *Of Microbes and Life*, edited by Jacques Monod and Ernest Borek, 56–61. New York: Columbia University Press, 1971.

Luria, S. E. *Life: The Unfinished Experiment*. New York: Scribner's, 1973.

Luria, S. E. "The Microbiologist and His Times." *Bacteriological Reviews* 32 (December 1968): 401–403.

Luria, S. E. "Modern Biology: A Terrifying Power." *The Nation*, October 20, 1969.

Luria, S. E. "Molecular Biology: Past, Present, and Future." *BioScience* 20 (December 15, 1970): 1289–1293, 1296.

Luria, S. E. "Mutations of Bacteria and of Bacteriophage." In *Phage and the Origins of Molecular Biology*, edited by John Cairns, Gunther Stent, and James D. Watson, 173–179. Cold Spring Harbor, NY: Cold Spring Harbor Laboratory Press, 1966, revised and expanded edition, 1992.

Luria, S. E. "Mutations of Bacterial Viruses Affecting Their Host Range." *Genetics* 30 (1945): 84–99.

Luria, S. E. "Phage; Colicins and Macroregulatory Phenomena." In *Nobel Lectures in Molecular Biology, 1933–1975*, edited by David Baltimore, 387–397. New York: Elsevier, 1977.

Luria, S. E. "Reactivation of Irradiated Bacteriophage by Transfer of Self-Reproducing Units." *Proceedings of the National Academy of Sciences* 33 (September 1947): 253–264.

Luria, S. E. "Recent Advances in Bacterial Genetics." *Bacteriological Reviews* 11 (1947): 1–40.

Luria, S. E. "Reflections on Democracy, Science, and Cancer." *Bulletin of the American Academy of Arts and Sciences* 30 (February 1977): 20–32.

Luria, S. E. "Reply." *New York Review of Books*, May 2, 1974.

Luria, S. E. "The Single Artificer." In *Responsible Science: The Impact of Technology on Society*, edited by Kevin B. Byrne, 49–71. San Francisco: Harper and Row, 1986.

Luria, S. E. "Slippery When Wet: Being an Essay on Science, Technology, and Responsibility." *Proceedings of the American Philosophical Society* 116 (October 1972): 351–356.

Luria, S. E. *A Slot Machine, a Broken Test Tube: An Autobiography*. New York: Harper Colophon, 1984.

Luria, S. E. "The T2 Mystery." *Scientific American*, April 1955.

Luria, S. E. "On Teaching Biology in a Biological Revolution: A Review of *Biology* by Helena Curtis." *Scientific American*, March 1969.

Luria, S. E. Testimony provided for February 11 appearance before the House Subcommittee on Science, Research and Technology, January 28, 1976, United States House of Representatives Correspondence Folder, Series I, SEL Papers, APS.

Luria, S. E. *36 Lectures in Biology*. Cambridge, MA: MIT Press, 1975.

Luria, S. E. "Viruses, Cancer Cells, and the Genetic Concept of Virus Infection." *Cancer Research* 20 (June 1960): 677–688.

Luria, S. E. "What Can Biologists Solve?" *New York Review of Books*, February 7, 1974.

Luria, S. E. and Thomas F. Anderson. "The Identification and Characterization of Bacteriophages with the Electron Microscope." *Proceedings of the National Academy of Sciences* 28 (1942): 127–130.

Luria, S. E., and James E. Darnell, Jr. *General Virology*. 2nd ed. New York: Wiley, 1967.

Luria, S. E., James E. Darnell, Jr., David Baltimore, and Allan Campbell. *General Virology*. 3rd ed. New York: Wiley, 1978.

Luria, S. E., and M. Delbrück. "Interference Between Inactivated Bacterial Virus and Active Virus of the Same Strain and of a Different Strain." *Archives of Biochemistry* 1 (1942): 207–218.

Luria, S. E., and M. Delbrück. "Mutations of Bacteria from Virus Sensitivity to Virus Resistance." *Genetics* 28 (1943): 491–511.

Luria, S. E., M. Delbrück, and T. F. Anderson. "Electron Microscope Studies of Bacterial Viruses." *Journal of Bacteriology* 46 (1943): 57–76.

Luria, S. E., and R. Dulbecco. "Genetic Recombinations Leading to Production of Active Bacteriophage from Ultraviolet Inactivated Bacteriophage Particles." *Genetics* 34 (1949): 93–125.

Luria, S. E., and F. M. Exner. "The Inactivation of Bacteriophages by X-rays—Influence of the Medium." *Proceedings of the National Academy of Sciences* 27 (1941): 370–375.

Luria, S. E., and F. M. Exner. "On the Enzymatic Hypothesis of Radiosensitivity." *Journal of the American Medical Association* 117 (1941): 2190.

Luria, S. E., Stephen Jay Gould, and Sam Singer. *A View of Life.* Menlo Park, CA: Benjamin/Cummings, 1981.

Luria, S. E., and Mary L. Human. "A Nonhereditary, Host-Induced Variation of Bacterial Viruses." *Journal of Bacteriology* 64 (1952): 557–569.

Luria, S. E., and Albert Szent-Györgyi. "Vietnam: A National Catastrophe." *Science* 158 (October 6, 1967): 47.

Lyons, Richard D. "Second H.E.W. Blacklist Includes Nobel Laureate: Institutes of Health Bar 48 More Researchers from Federal Panels." *New York Times,* October 20, 1969.

Lyons, Richard D. "Scientists Assail Bombing Policy." *New York Times,* December 29, 1972.

MacDougall, Curtis D. *Gideon's Army.* New York: Marzani and Munsell, 1965.

Macleod, Roy. "Science and Democracy: Historical Reflections on Present Discontents." *Minerva* 35 (1997): 369–384.

Mafai, Miriam. *Il Lungo Freddo: Storia di Bruno Pontecorvo, lo scienziato che scelse l'Urss.* Milano: Arnoldo Mondadori, 1992.

Magasanik, Boris. "A Charmed Life." *Annual Reviews in Microbiology* 48 (1994): 1–24.

Magasanik, Boris, John Ross, and Victor Weisskopf. "No Research Strike at M.I.T." *Science* 163 (February 7, 1969): 517.

Maier, Charles. *Recasting Bourgeois Europe: Stabilization in France, Germany and Italy in the Decade after World War I.* Princeton, NJ: Princeton University Press, 1975, 1988.

"March 4 at MIT." *New Republic* 160 (March 15, 1969): 10–11.

Markowitz, Norman D. *The Rise and Fall of the People's Century: Henry A. Wallace and American Liberalism, 1941–1948.* New York: Free Press, 1973.

Martin, Jerry L., and Anne D. Neal. "Defending Civilization: How Our Universities Are Failing America and What Can Be Done About It." A project of the Defense of Civilization Fund, American Council of Trustees and Alumni, November 11, 2001, revised and expanded February 2002.

Marx, Jean. "Cancer Cell Genes Linked to Viral *onc* Genes." *Science* 216 (May 14, 1982): 724.

Marx, Jean. "Oncogenes Reach a Milestone." *Science* 226 (December 23, 1994): 1942–1944.

Massachusetts Institute of Technology. "Annual Letter to Alumni and Friends of the Department of Biology, Massachusetts Institute of Technology." Numbers 8–10, 1960–1962.

Matusow, Allen J. *The Unraveling of America: A History of Liberalism in the 1960s*. New York: Harper Torchbooks, 1984.

May, Elaine Tyler. *Homeward Bound: American Families in the Cold War Era*. New York: Basic Books, 1988.

Mayr, Ernst, and William Provine, eds. *The Evolutionary Synthesis: Perspectives on the Unification of Biology*. Cambridge, MA: Harvard University Press, 1980.

McElheny, Victor E. "5 Nobel Men Back Him." *Boston Globe*, December 15, 1967.

McElheny, Victor E. *Watson and DNA: Making a Scientific Revolution*. Cambridge, MA: Perseus, 2003.

McElheny, Victor, and Herbert Black. "MIT Nobel Winner to Use Cash to End War." *Boston Globe*, October 16, 1969.

McKeon, Thomas J. "Gene Drives a Hard Biological Bargain." *Cleveland Press*, September 7, 1973.

Medvedev, Zhores. *The Rise and Fall of T. D. Lysenko*. New York: Anchor Books, 1971.

Mendelsohn, Everett I. "'Frankenstein at Harvard': The Public Politics of Recombinant DNA Research." In *Transformation and Tradition in the Sciences: Essays in Honor of I. Bernard Cohen*, edited by Everett Mendelsohn, 317–335. New York: Cambridge University Press, 1984.

Mendelsohn, Everett I. "The Politics of Pessimism: Science and Technology Circa 1968." In *Technology, Pessimism and Postmodernism*, Sociology of the Sciences Yearbook, 1993, edited by Yaron Ezrahi, Everett Mendelsohn, and Howard Segal, 151–173. Boston: Kluwer Academic Publishers, 1994.

Michaelis, Meir. *Mussolini and the Jews: German-Italian Relations and the Jewish Question in Italy, 1922–1945*. Oxford: Clarendon Press, for the Institute of Jewish Affairs, 1978.

Miller, James. *Democracy Is in the Streets: From Port Huron to the Siege of Chicago*. Cambridge, MA: Harvard University Press, 1994.

Miller, Judith, Stephen Engelberg, and William Broad. *Germs: Biological Weapons and America's Secret War*. New York: Simon & Schuster, 2001.

MIT Tech Talk. "MIT Honors Luria, Magasanik, Sharp." May 20, 1992.

Monod, Jacques. *Chance and Necessity: An Essay on the Natural Philosophy of Modern Biology*. Translated by Austryn Wainhouse. New York: Vintage Books, 1971.

Montagu, Ashley. *Man's Most Dangerous Myth: The Fallacy of Race*. New York: Columbia University Press, 1942. Reprinted in 1945, 1964, 1974, and 1977.

Moore, Kelly. *Disrupting Science: Social Movements, American Scientists and the Politics of the Military, 1945–1975*. Princeton, NJ: Princeton University Press, 2008.

Morange, Michel. *A History of Molecular Biology*. Translated by Matthew Cobb. Cambridge, MA: Harvard University Press, 1998.

Morowitz, Harold J. "A Man of Science Dissects and Catalogues His Life." *Boston Globe*, March 25, 1984.

Morrison, Philip. "Books." *Scientific American*, September 1984.

Morrison, Philip. "Review of *Life: The Unfinished Experiment*." *Scientific American*, August 1973.

Muller, Hermann J. "Variation Due to Change in the Individual Gene." *American Naturalist* 56 (1922): 32–50.

Mullins, Nicholas. "The Development of a Scientific Specialty: The Phage Group and the Origins of Molecular Biology." *Minerva* 10 (1972): 51–82.

Nation, The. "HUAC Eyes the Campus." December 12, 1966.

National Program for the Conquest of Cancer: Report of the National Panel of Consultants on the Conquest of Cancer authorized by S. Res. 376. Prepared for the Committee on Labor and Public Welfare, United States Senate. Washington, DC: United States Government Printing Office, 1971.

Needell, Allan A. *Science, Cold War and the American State: Lloyd V. Berkner and the Balance of Professional Ideals*. Washington, DC: Harwood Academic Publishers in association with the National Air and Space Museum, Smithsonian Institution, 2000.

Nelkin, Dorothy. *The University and Military Research: Moral Politics at M.I.T.* Ithaca, NY: Cornell University Press, 1972.

Nelson, Bryce. "HEW Security Checks Said to Bar Qualified Applicants from PHS." *Science* 165 (July 18, 1969): 269–271.

Nelson, Bryce. "HEW: Blacklists Scrapped in New Security Procedures." *Science* 167 (January 9, 1970): 154–156.

Nelson, Bryce. "Micro-Revolt of the Microbiologists over Detrick Tie." *Science* 160 (May 24, 1968): 862.

Nelson, Bryce. "M.I.T.'s March 4: Scientists Discuss Renouncing Military Research." *Science* 163 (March 14, 1969): 1175–1178.

Nelson, Bryce. "Scientists Increasingly Protest HEW Investigation of Advisors." *Science* 164 (June 27, 1969): 1499–1504.

Nelson, Bryce. "Scientists Plan Research Strike at M.I.T. on 4 March." *Science* 163 (January 24, 1969): 373.

Newsweek. "Ad-itorial Voices." November 6, 1967.

Newsweek. "Blacklist's Backlash." November 3, 1969.

New York Times. "An Appeal to President Kennedy and Premier Khrushchev." October 25, 1962.

New York Times. "An Appeal: We Must Tell the President." December 29, 1972.

New York Times. "Mr. President: STOP THE BOMBING." January 15, 1967, and January 22, 1967.

New York Times. "An Open Letter to President Kennedy: Why Are We Arming on Such a Scale?" August 21, 1962.

New York Times. "A Reply to Secretary Rusk on Vietnam." May 9, 1965.

New York Times. "Science Academy Adds 35 to Rolls." April 27, 1960.

New York Times. "67 Professors Back Policy in Vietnam." July 19, 1965.

New York Times. "Study of Life's Origin: MIT Research Will Center on Molecules, Cells, and Viruses." January 20, 1960.

New York Times. "Tie to Army Ended by Biology Society." May 9, 1968.

New York Times. "To the American People and to the Congress of the United States: An Open Letter on Vietnam." February 15, 1966.

New York Times. "To the President of the United States: STOP THE BLOODSHED." October 31, 1965.

New York Times. "2 Biologists Will Share $25,000 Research Prize." October 7, 1969.

Oakberg, Eugene F., and S. E. Luria, "Mutations to Sulfonamide Resistance in Staphylococcus-Aureus." *Genetics* 32 (1947): 249–261.

O'Brien, Tim. *The Things They Carried: A Work of Fiction*. New York: Penguin Books, 1991.

Olby, Robert. *Origins of Mendelism*. 2nd ed. Chicago: University of Chicago Press, 1985.

Olby, Robert. *The Path to the Double Helix: The Discovery of DNA*. Seattle: University of Washington Press, 1974, 2nd ed. 1994.

Olby, Robert. "From Physics to Biophysics." *History and Philosophy of the Life Sciences* 11 (1989): 305–309.

Olivo, Oliviero M. "Giuseppe Levi." In *Dictionary of Scientific Biography*, series edited by Charles C. Gillespie, 282–283. New York: Scribner's for the American Council of Learned Societies, 1973.

Olson, Clarence E. "The National Book Award Show." *St. Louis Post-Dispatch*, April 21, 1974.

O'Malley, Maureen A. "The Experimental Study of Bacterial Evolution and Its Implications for the Modern Synthesis of Evolutionary Biology." *Journal of the History of Biology* 51 (2018): 319–354.

O'Neill, William L. *Coming Apart: An Informal History of America in the 1960s*. Chicago: Quadrangle Books, 1971.

O'Neill, William L. *A Democracy at War: America's Fight at Home and Abroad in World War II*. New York: Free Press, 1993.

Orear, Jay, William F. Schreiber, Gerald Holton, Salvador E. Luria, Edwin E. Salpeter, Philip Morrison, Matthew Meselson, and Bernard T. Feld. "An Answer to Teller." *Saturday Evening Post*, April 14, 1962.

O'Reilly, Kenneth. *Hoover and the Unamericans: The FBI, HUAC, and the Red Menace*. Philadelphia: Temple University Press, 1983.

Paul, Diane B. "A War on Two Fronts: J. B. S. Haldane and the Response to Lysenkoism in Britain." *Journal of the History of Biology* 16 (1983): 1–37.

Pauling, Linus. "Appeal by American Scientists to the Government and Peoples of the World." *Bulletin of the Atomic Scientists* 13 (September 1957): 264–256.

Pauly, Philip J. *Biologists and the Promise of American Life: From Merriwether Lewis to Alfred Kinsey*. Princeton, NJ: Princeton University Press, 2000.

Pauly, Philip J. *Controlling Life: Jacques Loeb and the Engineering Ideal in Biology*. Berkeley: University of California Press, 1996.

Pave, Marvin. "Peace Groups Protest Kissinger Prize." *Boston Globe*, December 11, 1973.

Pells, Richard H. *The Liberal Mind in a Conservative Age: American Intellectuals in the 1940s and 1950s*. 2nd ed. Middletown, CT: Wesleyan University Press, 1985, 1989.

Pisanó, Giorgio. *Mussolini e gli Ebrei*. Milan: Edizioni FPE, 1967.

Podolsky, Scott. "The Role of the Virus in Origin-of-Life Theorizing." *Journal of the History of Biology* 29 (1996): 79–126.

Polenberg, Richard. *War and Society: The United States, 1941–1945.* Philadelphia: Lippincott, 1972.

Pollard, Ernest C. "Scientists' Views on Vietnam." *Science* 158 (October 27, 1967): 438–440.

Pollard, Ernest C. "Vietnam: Call for Scientific Help." *Science* 157 (August 18, 1967): 755–756.

Price, Don K. *America's Unwritten Constitution: Science, Religion, and Political Responsibility.* Cambridge, MA: Harvard University Press, 1985.

Primack, Joel, and Frank von Hippel. *Advice and Dissent: Scientists in the Political Arena.* New York: Basic Books, 1974.

Price, Don K. *The Scientific Estate.* Cambridge, MA: Belknap Press of Harvard University Press, 1965.

Prokasky, William F., William Palmer Taylor, and Robert B. Kelman. "Scientists' Views on Vietnam." *Science* 158 (October 27, 1967): 440–441.

Provine, William. *The Origins of Theoretical Population Genetics.* Chicago: University of Chicago Press, 1971.

Pugliese, Stanislao. *Carlo Rosselli: Socialist Heretic and Antifascist Exile.* Cambridge, MA: Harvard University Press, 1999.

Rabinowitch, Eugene. "The Purge of Genetics in the Soviet Union." *Bulletin of the Atomic Scientists* 5 (1949): 130.

Rai, Milan. *Chomsky's Politics.* New York: Verso, 1995.

Rankin-Hill, Lesley M., and Michael L. Blakey. "W. Montague Cobb: Physical Anthropologist, Anatomist and Activist." In *African-American Pioneers in Anthropology*, edited by Ira E. Harrison and Faye V. Harrison, 101–136. Chicago: University of Illinois Press, 1999.

Rasmussen, Nicolas. "The Mid-Century Biophysics Bubble: Hiroshima and the Biological Revolution in America, Revisited." *History of Science* 35 (1997): 245–293.

Rasmussen, Nicolas. *Picture Control: The Electron Microscope and the Transformation of Biology in America, 1940–1960.* Stanford, CA: Stanford University Press, 1997.

Reiner, John M. "Letter to the Editor." *Science* 158 (December 14, 1967): 1393.

Renzulli, Virgil. "Two Biophysicists Win Columbia's Horwitz Prize." *Columbia University Record* 22 (October 11, 1996). http://www.columbia.edu/cu/record/archives/vol22/vol22_iss6/record2206.16.html.

"Report of Annual Business Meeting, 3 May 1967." *ASM News* 33 (August 1967): 17.

"Report of the Annual Business Meeting, 8 May 1968." *ASM News* 34 (August 1968): 17.

"Report of the Committee Advisory to the U.S. Army Biological Laboratories." *ASM News* 33 (August 1967): 20–21.

"Report of the Council and Council Policy Committee, 29 April–2 May, 1967." *ASM News* 33 (August 1967): 12.

"Report of the Council and CPC." *ASM News* 34 (August 1968): 14.

Rettig, Richard A. *Cancer Crusade: The Story of the National Cancer Act of 1971.* Princeton, NJ: Princeton University Press, 1977.

Rheinberger, Hans-Jorg. *Toward a History of Epistemic Things: Synthesizing Proteins in the Test Tube.* Stanford, CA: Stanford University Press, 1997.

Ribatti, Domenico. "Tre Compagni di Studi. Gli anni torinesi di Renato Dulbecco, Rita Levi-Montalcini e Salvador Luria." *Rivista di Storia della Medicina* Anno III NS (24) fasc. 2 (Juglio-Dicembre 1993): 43–53.

Rosenberg, Charles E. *No Other Gods: On Science and American Social Thought.* Baltimore: Johns Hopkins University Press, 1997.

Rosenblum, S., and S. E. Luria. "Fernand Holweck, 1889–1941." *Science* 96 (October 9, 1942): 329–330.

Ross, Walter S. *Crusade: The Official History of the American Cancer Society.* New York: Arbor House, 1987.

Rossiter, Margaret W. *Women Scientists in America: Before Affirmative Action, 1940–1972.* Baltimore: Johns Hopkins University Press, 1995.

Roth, Cecil. *The History of the Jews of Italy.* Philadelphia: Jewish Publication Society of America, 1946.

Rudolph, John L. *Scientists in the Classroom: The Cold War Reconstruction of American Science Education.* New York: Palgrave, 2002.

Rusk, Dean. "The Control of Force in International Relations." *Department of State Bulletin* 52 (May 10, 1965): 694–701.

Russell, Nicholas. "Toward a History of Biology in the Twentieth Century: Directed Autobiographies as Historical Sources." *British Journal of the History of Science* 21 (1988): 77–89.

Saladino, Salvatore. *Italy from Unification to 1919: Growth and Decay of a Liberal Regime.* Arlington Heights, IL: Harlan Davidson, 1970.

Salloch, Roger. "Cambridge: March 4, the Movement, and M.I.T." *Bulletin of the Atomic Scientists* 25 (May 1969): 32–35.

Salomon, Jean-Jacques. "Science, Technology and Democracy." *Minerva* 38 (2000): 33–51.

Sankaran, Neeraja. "When Viruses Were Not in Style: Parallels in the History of Chicken Sarcoma Viruses and Bacteriophages." *Studies in History and Philosophy of Biological and Biomedical Sciences* 48 (2014): 189–199.

Sapp, Jan. "The Bacterium's Place in Nature." In *Microbial Phylogeny and Evolution: Concepts and Controversies*, edited by Jan Sapp, 3–52. New York: Oxford University Press, 2005.

Sapp, Jan. *Beyond the Gene: Cytoplasmic Inheritance and the Struggle for Authority in Genetics*. New York: Oxford University Press, 1987.

Sapp, Jan. *Where the Truth Lies: Franz Moewus and the Origins of Molecular Biology*. New York: Cambridge University Press, 1990.

Sarkar, Sohatra. "Lamark *contre* Darwin, Reduction *versus* Statistics: Conceptual Issues in the Controversy over Directed Mutagenesis in Bacteri." In *Organisms and the Origins of Self*, Boston Studies in the Philosophy of Science 129, edited by Alfred I. Tauber, 235–271. Boston: Kluwer Academic Publishers, 1991.

Sarti, Roland. *Fascism and the Industrial Leadership in Italy, 1919–1940: A Study in the Expansion of Private Power under Fascism*. Berkeley: University of California Press, 1971.

Sayre, Anne. *Rosalind Franklin and DNA*. New York: W. W. Norton, 1975.

Scarpelli, Giacomo, ed. *Storia della biologia in Italia*. Rome: Omaggio Del Gruppo Ferruzzi, 1988.

Schaffer, Simon. "Late Victorian Metrology and Its Instrumentation: A Manufactory of Ohms." In *Invisible Connections: Instruments, Institutions and Science*, edited by Robert Bud and Susan Cozzens, 23–56. Bellingham, WA: SPIE Optical Engineering Press, 1991.

Schalk, David L. *War and the Ivory Tower: Algeria and Vietnam*. New York: Oxford University Press, 1991.

Scheffler, Robin Wolfe. *A Contagious Cause: The American Hunt for Cancer Viruses and the Rise of Molecular Medicine*. Chicago: University of Chicago Press, 2019.

Schmidt, Mark Ray, ed. *The 1970s*. America's Decades Series. San Diego: Greenhaven, 2000.

Schoenfeld, Lowell. "Anti-Nepotism Rules." *AUUP Bulletin* 48 (Summer 1962): 187–188.

Schorske, Carl. *The Age of McCarthyism: A Brief History with Documents*. Boston: Bedford Books, 1994.

Schorske, Carl. *Fin-De-Siecle Vienna: Politics and Culture*. New York: Vintage Books, 1981.

Schrecker, Ellen. *Many Are the Crimes: McCarthyism in America*. Princeton, NJ: Princeton University Press, 1999.

Schrecker, Ellen. *No Ivory Tower: McCarthyism and the Universities*. New York: Oxford University Press, 1986.

Schulman, Bruce J. *The Seventies: The Great Shift in American Culture, Society and Politics*. New York: Free Press, 2001.

Schweber, S. S. *In the Shadow of the Bomb: Bethe, Oppenheimer, and the Moral Responsibility of the Scientist*. Princeton, NJ: Princeton University Press, 2000.

Segre, Dan Vittorio. *Memoirs of a Fortunate Jew: An Italian Story*. Bethesda, MD: Adler and Adler, 1987.

Segré, Emilio. "Enrico Fermi." In *Dictionary of Scientific Biography*, series edited by Charles C. Gillespie, 576–584. New York: Scribner's for the American Council of Learned Societies, 1971.

Segré, Emilio. *Enrico Fermi: Physicist*. Chicago: University of Chicago Press, 1970.

Selya, Rena. "Defending Scientific Freedom and Democracy: The Genetics Society of America's Response to Lysenko." *Journal of the History of Biology* 45 (Fall 2012): 415–442.

Severo, Richard. "Salvador E. Luria Is Dead at 78; Shared Nobel Prize in Medicine." *New York Times*, February 7, 1991.

Sharp, Phillip A. "Jim as a Mentor, 1971–1974." In *Inspiring Science: Jim Watson and the Age of DNA*, edited by John R. Inglis, Joseph Sambrook, and Jan A. Witkowski, 273–276. Cold Spring Harbor, NY: Cold Spring Harbor Laboratory Press, 2003.

Sharp, Phillip A. "Life Sciences at MIT: A History and Perspective." *MIT Faculty News* 18 (January/February 2006).

Shelter, Stanwyn G. "Botany—A Passing Phase?" *AIBS Bulletin* 13 (December 1963): 23–25.

Sherry, Michael S. *In the Shadow of War: The United States Since the 1930s*. New Haven, CT: Yale University Press, 1995.

Shils, Edward. "Scientists and the Antiscience Movement." Foreword to *Storm over Biology: Essays on Science, Sentiment and Public Policy*, by Bernard D. Davis. Buffalo, NY: Prometheus Books, 1986.

Shortland, Michael. "Exemplary Lives: A Study of Scientific Autobiographies." *Science and Public Policy* 15 (June 1988): 170–179.

Shortland, Michael, and Richard Yeo, eds. *Telling Lives in Science: Essays on Scientific Biography*. Cambridge: Cambridge University Press, 1996.

Sidik, Saima. "The Untidy Experiment that Catalyzed Recombinant DNA Technology." MIT Department of Biology News, December 15, 2020. https://biology.mit.edu /news/untidy-experiment-catalyzed-recombinant-dna-tech/.

Silver, Simon. "Review of *A Slot Machine, a Broken Test Tube: An Autobiography*." *ASM News* 51 (September 1985): 496.

Simpson, George Gaylord. *Biology and Man*. New York: Harcourt, Brace & World, 1969.

Sizer, Irwin. "Life Sciences–M.I.T. Style." *Technology Review* 59 (June 1957): 423–426, 440.

Sloan, Philip R. "Molecularizing Chicago—1945–1965: The Rise, Fall, and Rebirth of the University of Chicago Biophysics Program." *Historical Studies in the Natural Sciences* 44 (2014): 364–412.

Small, Melvin, and William D. Hoover, eds. *Give Peace a Chance: Exploring the Vietnam Antiwar Movement*. Syracuse, NY: Syracuse University Press, 1992.

Smigel, Kara. "1971–1991: Virus Cancer Research Pays Rich Dividends." *Journal of the National Cancer Institute* 83 (November 6, 1991): 1528–1531.

Smith, Alice Kimball. *A Peril and a Hope: The Scientists' Movement in America, 1945–47*. Chicago: University of Chicago Press, 1965.

Smith, Crosbie, and M. Norton Wise. *Energy and Empire: A Biographical Study of Lord Kelvin*. New York: Cambridge University Press, 1989.

Smith, M. L. "No Tears for Luria." *Boston Herald Traveler*, October 1969.

Smocovitis, Vassiliki Betty. "Organizing Evolution: Founding the Society for the Study of Evolution (1939–1950)" *Journal of the History of Biology* 27 (1994): 241–309.

Smocovitis, Vassiliki Betty. *Unifying Biology: The Evolutionary Synthesis and Evolutionary Biology*. Princeton, NJ: Princeton University Press, 1996.

Söderqvist, Thomas, ed. *The Historiography of Contemporary Science and Technology*. Amsterdam: Harwood Academic Press, 1997.

Söderqvist, Thomas, ed. *The History and Poetics of Scientific Biography*. London: Ashgate, 2007.

Söderqvist, Thomas. *Science as Autobiography: The Troubled Life of Niels Jerne*. Translated by David Mel Paul. New Haven, CT: Yale University Press, 2003.

Sonneborn, T. M., ed. *The Control of Human Heredity and Evolution*. New York: Macmillan, 1965.

Spackman, Barbara. *Fascist Virilities: Rhetoric, Ideology and Social Fantasy in Italy*. Minneapolis: University of Minnesota Press, 1996.

Sperling, Susan. "Ashley Montagu (1905–1999)." *American Anthropologist* 102 (September 2000): 583–588.

Sprinkle, Robert Hunt. *Profession of Conscience: The Making and Meaning of Life-Sciences Liberalism*. Princeton, NJ: Princeton University Press, 1994.

Stahl, Franklin W., ed. *We Can Sleep Later: Alfred D. Hershey and the Origins of Molecular Biology*. Cold Spring Harbor, NY: Cold Spring Harbor Laboratory Press, 2000.

Star, S. L., and J. R. Griesemer. "Institutional Ecology, 'Translations,' and Boundary Objects: Amateurs and Professionals in Berkeley's Museum of Vertebrate Zoology, 1907–39." *Social Studies of Science* 19 (1989): 387–420.

"Statement of Dr. Salvador E. Luria, Professor of Biology and Institute Professor, Massachusetts Institute of Technology." Tuesday, September 28, 1971. Published in *Part 2 National Cancer Act of 1971, Hearings before the Subcommittee on Public Health and the Environment of the Committee on Interstate and Foreign Commerce, House of Representatives*. Serial No. 92–41. Washington, DC: U.S. Government Printing Office, 1971, 485–490.

Stent, Gunther S. *The Coming of the Golden Age: A View of the End of Progress*. Garden City, NY: Natural History Press, 1969.

Stent, Gunther S. "Genes to Cells to Complex Organisms to Species." *New York Times Book Review*, September 8, 1974.

Stent, Gunther S. "Max Delbrück, 1906–1981." *Genetics* 101 (May 1982): 1–16.

Stent, Gunther S. *Molecular Genetics: An Introductory Narrative*. San Francisco: W. H. Freeman, 1970.

Stent, Gunther S. *Nazis, Women and Molecular Biology: Memoirs of a Lucky Self-Hater*. Kensington, CA: Briones Books, 1998.

Stent, Gunther S. "The 1969 Nobel Prize for Physiology or Medicine." *Science* 166 (October 24, 1969): 479–481.

Stent, Gunther S. "That Was the Molecular Biology That Was." *Science* 160 (April 26, 1968): 390–395.

Stille, Alexander. *Benevolence and Betrayal: Five Italian Jewish Families under Fascism*. New York: Summit Books, 1991.

Stoddard, George D. *The Pursuit of Education: An Autobiography*. New York: Vintage, 1981.

Studer, Kenneth E., and Daryl E. Chubin. *The Cancer Mission: Social Contexts of Biomedical Research*. Sage Library of Social Research Volume 103. London: Sage, 1980.

Sturtevant, A. H. *A History of Genetics*. New York: Harper and Row, 1965.

Sulloway, Frank. *Freud, Biologist of the Mind: Beyond the Psychoanalytic Legend*. Cambridge, MA: Harvard University Press, 1992.

"Summary of Testimony of Linus Pauling." *Bulletin of the Atomic Scientists* 12 (January 1956): 28.

Summers, William C. *Félix d'Herelle and the Origins of Molecular Biology*. New Haven, CT: Yale University Press, 1999.

Summers, William C. "Felix Hubert d'Herelle (1873–1949): History of Scientific Mind." *Bacteriophage* 6 (2016): e1270090.

Summers, William C. "How Bacteriophage Came to Be Used by the Phage Group." *Journal of the History of Biology* 26 (1993): 255–267.

Summers, William C. "Inventing Viruses." *Annual Review of Virology* (2014): 25–35.

Susman, Millard. "The Cold Spring Harbor Phage Course (1945–1970): A 50th Anniversary Remembrance." *Genetics* 139 (1995): 1101–1106.

Swearingen, Rodger. "The Vietnam Critics in Perspective." *Communist Affairs* 4 (June 1966): 7–9.

Teich, Albert H., ed. *Scientists and Public Affairs*. Cambridge, MA: MIT Press, 1974.

Teller, Edward. *The Legacy of Hiroshima*. New York: Doubleday, 1962.

Thackray, Arnold, ed. *Science after '40: Osiris* 7. Chicago: University of Chicago Press, 1992.

Theoharis, Athan, ed. *Beyond the Hiss Case: The FBI, Congress, and the Cold War*. Philadelphia: Temple University Press, 1982.

Theoharis, Athan. *Spying on Americans: Political Surveillance from Hoover to the Huston Plan*. Philadelphia: Temple University Press, 1978.

Time. "A Policy of Protest." February 28, 1969.

Todd, Richard. "The 'Ins' and 'Outs' at M.I.T." *New York Times Magazine*, May 18, 1969.

Tomes, Robert. *Apocalypse Then: American Intellectuals and the Vietnam War, 1954–1975*. New York: New York University Press, 1998.

Travis, John. "In Stockholm, a Clean Sweep for North America, Medicine: Discovery of Genes in Pieces Wins for Two Biologists." *Science* 262 (October 22, 1993): 506.

"T.R.B. from Washington: Fellow-Travelers and HUAC Hellzapoppin." *New Republic* 155 (August 27, 1966): 4–6.

Tuchman, Barbara. "1976: Bicentennial Reflections." *Newsweek*, July 12, 1976.

Tuttle, William. *"Daddy's Gone to War": The Second World War in the Lives of American Children*. New York: Oxford University Press, 1993.

Twort, Antony. *In Focus, Out of Step: A Biography of Frederick William Twort F.R.S., 1877–1950*. Dover, NH: Alan Sutton, 1993.

Twort, F. W. "An Investigation of the Nature of Ultra-Microscopic Viruses." *Lancet* 2 (1915): 1241–1243.

Ueda, Reed. "The Changing Path to Citizenship: Ethnicity and Naturalization During World War II." In *The War in American Culture: Society and Consciousness During World*

War II, edited by Lewis A. Erenberg and Susan E. Hirch, 202–216. Chicago: University of Chicago Press, 1996.

U.S. News and World Report. "Antiwar Protests: A Weapon for Communism." November 13, 1967.

van Helvoort, Ton. "The Construction of Bacteriophage as Bacterial Virus: Linking Endogenous and Exogenous Thought Styles." *Journal of the History of Biology* 27 (1994): 91–139.

van Helvoort, Ton. "History of Virus Research in the Twentieth Century: The Problem of Conceptual Continuity." *History of Science* 32 (1994): 185–235.

van Helvoort, Ton, and Neeraja Sankaran. "How Seeing Became Knowing: The Role of the Electron Microscope in Shaping the Modern Definition of Viruses." *Journal of the History of Biology* 52 (2019): 125–160.

Verschuuren, Gerard M. *Life Scientists: Their Convictions, Their Activities, and Their Values.* North Andover, MA: Genesis, 1995.

Visconti, N. "Mating Theory." In *Phage and the Origins of Molecular Biology*, edited by John Cairns, Gunther Stent, and James D. Watson, 142–149. Cold Spring Harbor, NY: Cold Spring Harbor Laboratory Press, 1966, revised and expanded edition, 1992.

von Hippel, Frank. *Citizen Scientist.* New York: American Institute of Physics, 1991.

Wade, Nicholas. "Salk Institute: Elitist Pursuit of Biology with a Conscience." *Science* 178 (November 24, 1972): 846–849.

Walton, Richard J. *Henry Wallace, Harry Truman, and the Cold War.* New York: Viking, 1976.

Wang, Jessica. *American Science in an Age of Anxiety: Scientists, Anticommunism and the Cold War.* Chapel Hill: University of North Carolina Press, 1999.

Wang, Zuoyue. *In Sputnik's Shadow: The President's Science Advisory Committee and Cold War America.* New Brunswick, NJ: Rutgers University Press, 2008.

Washington Post. "An Open Letter to the American People: A Count of Broken Promises." July 7, 1966.

Waterson, A. P., and Lise Wilkinson. *An Introduction to the History of Virology.* New York: Cambridge University Press, 1978.

Watson, James D. "Director's Report" from *Cold Spring Harbor Laboratory 1989 Annual Report.* In *Inspiring Science: Jim Watson and the Age of DNA*, edited by John R. Inglis, Joseph Sambrook, and Jan A. Witkowski, 263–272. Cold Spring Harbor, NY: Cold Spring Harbor Laboratory Press, 2003.

Watson, James D. *The Double Helix: A Personal Account of the Discovery of the Structure of DNA.* Norton Critical Edition, edited by Gunther Stent. New York: W. W. Norton, 1980.

Watson, James D. *Genes, Girls, and Gamow*. New York: Oxford University Press, 2001.

Watson, James D. "Growing Up in the Phage Group." In *Phage and the Origins of Molecular Biology*, edited by John Cairns, Gunther Stent, and James D. Watson, 239–245. Cold Spring Harbor, NY: Cold Spring Harbor Laboratory Press, 1966, revised and expanded edition, 1992.

Watson, James D. "The Human Genome Project: Past, Present, and Future." *Science* 248 (April 6, 1990): 44–49.

Watson, James D. *A Passion for DNA: Genes, Genomes and Society*. Cold Spring Harbor, NY: Cold Spring Harbor Laboratory Press, 2000.

Watson, James D. "Salvador E. Luria (1912–1991)." *Nature* 350 (March 14, 1991): 113.

Watson, James D., and John Tooze, eds. *The DNA Story: A Documentary History of Gene Cloning*. San Francisco: W. H. Freeman, 1981.

Weinberg, Alvin M. "The Federal Laboratories and Science Education." *Science* 136 (April 6, 1962): 27–30.

Weiner, Jonathan. *Time, Love, Memory: A Great Biologist and His Quest for the Origins of Behavior*. New York: Vintage Books, 1999.

Weisskopf, Victor. *The Joy of Insight: Passions of a Physicist*. New York: Basic Books, 1991.

Well, Tom. *The War Within: America's Battle over Vietnam*. Los Angeles: University of California Press, 1994.

Wellerstein, Alex. "The Demon Core and the Strange Death of Louis Slotin." *New Yorker*, May 21, 2016. https://www.newyorker.com/tech/annals-of-technology/demon-core-the-strange-death-of-louis-slotin.

White, Theodore H. *Breach of Faith: The Fall of Richard Nixon*. New York: Atheneum, 1975.

Whitfield, Stephen J. *The Culture of the Cold War*. Baltimore: Johns Hopkins University Press, 1991.

Wilkinson, James. *The Intellectual Resistance in Europe*. Cambridge, MA: Harvard University Press, 1981.

Willey, Malcom W. *Depression, Recovery and Higher Education: A Report by Committee Y of the American Association of University Professors*. New York: McGraw-Hill, 1937.

Wilmut, Ian, Keith Campbell, and Colin Tudge. *The Second Creation: Dolly and the Age of Biological Control*. New York: Farrar, Straus and Giroux, 2000.

Wilson, James Q. "Lock 'Em Up and Other Thoughts on Crime." *New York Times Magazine*, March 9, 1975.

Wilson, James Q. "Reply." *New York Times Magazine*, April 13, 1975.

Wise, M. Norton. "Mediating Machines." *Science in Context* 2 (1988): 77–113.

Wise, M. Norton. "Work and Waste: Political Economy and Natural Philosophy in 19th Century Britain." *History of Science* 27 (1989): 263–301, 391–449; 28 (1990): 221–261.

Wiseman, Steven R. "World Books Presents Its Oscars." *New York Times*, April 19, 1974.

Witkin, Evelyn M. "Chances and Choices: Cold Spring Harbor 1944–1955." *Annual Review of Microbiology* 56 (2002): 1–15.

Witkin, Evelyn M. "Genetics of Resistance to Radiation in *Escherichia coli*." *Genetics* 32 (1947): 221–248.

Witkowski, Jan, ed. *The Inside Story: DNA to RNA to Protein*. Cold Spring Harbor, NY: Cold Spring Harbor Laboratory Press, 2005.

Wolfe, Audra. "Biology and Liberty for All: The Biological Sciences Curriculum Study." Paper presented at the History of Science Society meeting, Denver, CO, November 9, 2001.

Wolfe, Audra. "The Cold War Context of the Golden Jubilee, Or, Why We Think of Mendel as the Father of Genetics." *Journal of the History of Biology* 45 (Fall 2012): 389–414.

Wolfe, Audra. *Freedom's Laboratory: The Cold War Struggle for the Soul of Science*. Baltimore: Johns Hopkins University Press, 2019.

Wolfe, Audra. "Germs in Space: Joshua Lederberg, Exobiology, and the Public Imagination, 1958–1964." *Isis* 93 (2002): 183–205.

Wolfe, Audra. *Speaking for Nature and Nation: American Biologists as Cold War Intellectuals, 1947–1972*. Ph.D. diss., University of Pennsylvania, 2002.

Wolfe, Audra. "What Does It Mean to Go Public? The American Response to Lysenko, Reconsidered." *Historical Studies in the Natural Sciences* 40 (2010): 48–78.

Wolfle, Dael. "AAAS Council Meeting, 1955." *Science* 123 (February 17, 1956): 268–270.

Wolfle, Dael. *Renewing a Scientific Society: The American Association for the Advancement of Science from World War II to 1970*. Washington, DC: American Association for the Advancement of Science, 1989.

Wollman, E., F. Holweck, and S. Luria. "Effects of Radiations on Bacteriophage C16." *Nature* 145 (1940): 935.

Woodward, C. Vann. "The Aging of America." *American Historical Review* 82 (June 1977): 583–594.

Wright, Susan. *Molecular Politics: Developing American and British Regulatory Policy for Genetic Engineering, 1972–1982*. Chicago: University of Chicago Press, 1994.

Wynne, E. Staten. "Letter to the Editor." *Science* 158 (December 14, 1967): 1393.

Yoxen, E. J. "Where Does Schrödinger's *What Is Life?* Belong in the History of Molecular Biology?" *History of Science* 27 (1979): 17–52.

Zachary, G. Pascal. *Endless Frontier: Vannevar Bush, Engineer of the American Century.* New York: Free Press, 1997.

Zaroulis, Nancy, and Gerald Sullivan. *Who Spoke Up? American Protest Against the War in Vietnam, 1963–1975.* New York: Doubleday, 1984.

Zeiler, Thomas W. *Dean Rusk: Defending the American Mission Abroad.* Wilmington, DE: Scholarly Resources, 2000.

Zuccotti, Susan. *The Italians and the Holocaust: Persecution, Rescue and Survival.* Lincoln: University of Nebraska Press, 1987, reprinted 1996.

INDEX